Chemistry of Fuel, Furnaces and d-Block Elements

Chemistry of Fuel, Furnaces and d-Block Elements

Manish Rathi

CYBER TECH PUBLICATIONS
4264/3, Ansari Road, Daryaganj, New Delhi-110002 (India)
Ph.: 011-23244078, 011-43559448 Fax: 011-23280028
E-Mail: cyberpublicationsdelhi@yahoo.com
Website: www.cybertechpublications.com

CHEMISTRY OF FUEL, FURNACES AND D-BLOCK ELEMENTS

Manish Rathi

First Edition 2010

Published by : Asian Offset Printers

G.S. Rawat for **Cyber Tech Publications**
4264/3, Ansari Road, Daryaganj, New Delhi-110002 (India)
Ph.: 011-23244078, 011-43559448 Fax: 011-23280028
E-Mail: cyberpublicationsdelhi@yahoo.com
Website: www.cybertechpublications.com

PREFACE

This books has been specially written to the requirements syllabus of the students of all Indian Universities. The Subject matter has been discussed in such a simple way that the students will find no difficulty to understand it. Each chapter of this book contains complete theory and fairly large number of solved examples. Sufficient problems has also been selected from various universities examination papers.

The proofs of various principles and examples has been given with minute details. The subject matter has been presented in a simple and systematic manner. To familiarize the student about expected type of questions in the examinations, important summary and review questions have been provided at the end of each chapter.

Suggestions for further improvement of this book are most welcome and would be incorporated in the next edition. The author will feel amply rewarded if the book serve the purpose for which it is meant.

— *Author*

PREFACE

Contents

Pages

1

Basic Concept of Fuels and Furnaces

INTRODUCTION

Nuclear energy may also be called source of heat energy. In simple words any fissionable material will be termed as fuel. Electric energy when used as a source of heat is also called fuel. Chemical compounds are also fuels. For example, in rockets, alcohol, ammonia and hydrazine are used as fuels. *"Any chemical or reactant which produces energy in a form that can be used for producing power is called fuel"*. From the above examples it is evident that combustion is not necessary for a fuel to give out energy. Thus, the modern definition of a fuel is;

MINERAL WEALTH

The term *mineral* is a general one and is applied to a pure homogeneous inorganic substance found in nature. Minerals from which metals can be economically extracted are called ores. Ores often occur associated with worthless rocky material, called the *gangue or matrix*. The natural mode of occurrence of an element largely depends upon its chemical nature and particularly on its electrode potential, size, charge and polarizability which in turn depend on the position of the element in the periodic table. The earth's crust may be regarded as the biggest source of metals whereas some salts are also found in sea. Metals occur in nature, sometimes free but mostly in combined state. Those found in free or native state are copper, silver, gold, mercury and platinum. Others are found in the combined state in the earth and called minerals.

The economic position of any country is more or less determined by the Mineral Wealth of the country. If a country is rich in Mineral Wealth, it is definitely more prosperous. A living example is U.S.A which is regarded as the richest country and this is due to its large deposits of minerals. Thus, one can safely say that *Mineral wealth of a country determines the economic position of a country.*

India is quite fortunate to have been gifted by nature with rich mineral resources especially of coal, mica, iron, manganese, aluminium, magnesium, titanium, rare earths, thorium and several other atomic minerals. The most extensively worked minerals are coal, mica, manganese, iron, gypsum, limestone and bauxite.

As far as the non-ferrous metals such as copper, zinc and lead are concerned India at present depends upon foreign imports but deposits of potential value are being explored.

India was famous in the past for its Mineral Wealth such as gold, silver and diamonds. This invited several invasions from some Asian and European countries. That is why Britishers were not interested in leaving this country.

We Indians feel proud in declaring that use of minerals like gold, mercury, iron and black mica in medicines was first introduced by us.

POSITION OF INDIA

India possesses abundant reserves of coal, mica, iron ore, manganese ore, ores of aluminium, magnesium and titanium, high grade refractories, rare earths, thorium and other atomic minerals. Mainly, these minerals are found in Bihar, Orissa, West Bengal, Madhya Pradesh, Rajasthan, Mysore, Andhra Pradesh and Goa.

We are trying to summarise an account of important minerals giving particulars of the occurrence, reserves, quality, recent out-put and industrial uses.

1. **Apatite and Rock Phosphate :** As a result of recent investigations a number of phosphorite deposits of considerable magnitude, have been located in states of Rajasthan and Uttar Pradesh, of which the one located at *Thamar Kotra in Udaipur District* is the largest discovered so far. The reserves in Thaijar Kotra deposits are estimated at more than 40 million tons with 20-30% of P_2O_5. Mining in this area commenced id 1969. Small deposits of apatite are also found in Bihar, Orissa and Andhra

Pradesh. In Bihar reserves of 1.09 million tons of apatite with 16.5 per cent P_2O_5, have been estimated.

2. **Asbestos** : In India, asbestos is mainly found in Andhra Pradesh, Bihar and Mysore. Good quality of asbestos is mainly obtained from Andhra Pradesh. Asbestos is widely used in the manufacture of special types of cements, for heat insulation and lining in the automobile industry.

3. **Barytes :** Barytes is the principal barium mineral. Deposits of this are mainly found in Andhra Pradesh, Rajasthan and Madras. It is very useful substance and is used in paints, as filler in rubber, textiles, paper, leather, etc. It is also used in the ceramic and chemical industries.

4. **Bauxite :** The occurrence of bauxite is widespread in the country. Workable deposits are found in Bihar, Madhya Pradesh, Tamil Nadu, Maharashtra, Mysore, Orrisa, Goa, Gujarat, Uttar Pradesh, and Jammu and Kashmir. The total reserves in all the states of all grades are of the order of 236 million tonnes of which are the reserves of high grade containing 51%, Al_2O_3 and above are placed at 116 million tons. India exports bauxite mainly to Japan, Italy, West Germany and Australia.

5. **Beryl (2BeO. Al_2O_3.6SiO_2) :** It is used in atomic reactor as a moderator. It is mainly found in Rajasthan, Bihar and Andhra Pradesh. Beryl is also the chief source of Beryllium.

6. **Chromite (FeO. Cr_2O_3) :** It is the chief ore of chromium. It occurs mainly in Orissa, Bihar, Mysore and Maharastra. Occurrence has also been reported from Tamil Nadu and Andhra Pradesh. The state of Orissa is the principal producer for over 90% of the total chromite production. The total reserves of chromite are placed from 15.5 to 15.7 million tonnes.

   Chromite is widely used in producing chromium-iron alloys, refractories and chemicals like sodium and potassium dichromates.

7. **Coal and Lignite :** The most important coal fields are at Raniganj in West Bengal and Bokaro in Bihar. The best coal comes from Giridh, Jharia and Rani Ganj. Other states in which coal is found are Madhya Pradesh, Tamil Nadu, Maharastra, Assam, Rajasthan and Kashmir. The proved and inferred reserves of all types are

estimated at 167.96 and 130.782 million tonnes respectively by Geological Survey of India. Burma and Ceylon were the chief importer of coal. Lignite occurs at Patna and Khari in Bikaner district of Rajasthan, Jammu and Kashmir, Kutch and Broach districts in Gujarat. The multipurpose 'South Arcot Lignite Project' at Neyveli is planned to increase the mine out put to 6 million tones. Coal is mainly exported to Japan and Ceylon.

8. **Copper :** There are three districts copper belts in the country, viz., Singhbum in Bihar, Khetri, and Darbia in Rajasthan and Agnigundal in Andhra Pradesh. There are several other deposits known In Tamil Nadu, Maharastra, Mysore and Gujarat which are under investigation. In terms of copper metal, the total reserve is estimated to be about 3.49 million tonnes. These days, copper is being imported.

9. **Corundum :** It is found in Assam, Madhya Pradesh and Mysore. The reserves of high grade corundum of Pipra in M.P., Some Pahar in Assam are of the order 100000 and 80000 tonnes respectively.

   At pipra, it is estimated to be the order of 4 lakh tonnes. Due to its hardness (next to diamond) it is mainly used in the manufacture of abrasives.

10. **Diamonds :** Occurrence of diamonds in India is reported mainly from Anantpur district, Batulpadu in Krishana district and in Krishana river in Andhra Pradesh, Panna and Hinota areas in the Panna district and Angor in Chattarpur district of Madhya Pradesh. Potential diamond reserves of the Mayghawan pipe (M.P) have been found to be about 3 million carats.

11. **Dolomite :** Dolomite occurs in West Bengal, Orissa, Madhya Pradesh and Uttar Pradesh. The approved deposites of dolomite in India are of the order 50 million tonnes. Dolomite is a mineral source of magnesium and calcium.

    It is used in the manufacture of iron and steel, ferromanganese, ferrosilicon and refractories for ceramics, rubber and tanneries.

12. **Fluorspar :** It is the mineral source of calcium. It is found mainly in Gujarat, Madhya Pradesh and Rajasthan. It is used as an important fluxing mineral in iron, steel and aluminium industry.

    According to the Geological Survey of India, the inferred reserves of fluorspar in the Amba Dungar deposit are estimated to be of the order 11.6 million tonnes containing 30 per cent calcium fluoride. In Chandi Durg area of Durg district deposits have been estimated at about 3,90.000 tonnes containing 22 to 39% CaF_2 content.

13. **Galena :** It is the principal source of lead. The Zawar mines of Udaipur district of Rajasthan are the only producing source of lead and zinc ore in the country. Reserves of this mineral in this area are estimated to be about 100 million tonnes, lead deposits have also been discovered in Lokedeya Sargipalli in Sundergarh district, Orissa and reserves of 0.12 million tonnes containing 6% lead have been estimated.

14. **Gold :** The entire production of gold is obtained from Mysore state. Kolar and Hutti are the two mining fields. The total reserves in these mines about were 45639 kgms. Nearly all the gold was retained by Government of India to provide stability to our paper currency. In 1994, these mines have been abandoned.

15. **Graphite :** It is an allotropic form of carbon and is mainly used in the atomic reactor as a moderator, in electrical furnaces, in lubricants and in the manufacture of special types of crucibles. It is found mainly in Orissa, Bihar, Madhya Pradesh and Andhra Pradesh.

16. **Gypsum :** It is calcium sulphate with two molecules of water of crystallisation and is used in the manufacture of fertilizer, cement, insecticides, plaster of pans, etc. Rajasthan contains the largest reserves of 1130 million tonnes of gypsum out of the country's total reserves of 1,189 million tonnes. Other states having gypsum deposits are Tamil Nadu, Uttar Pradesh, Himachal Pradesh, Gujarat, Maharashtra, Kashmir and Mysore. Except Mysore, all the states mentioned above produce gypsum.

17. **Ilmenite :** It is the principal ore of titanium and is chiefly used for the production of white pigment. However, it is exported to U.K. and U.S.A. It occurs mainly in the eastern and western coast of India associated with the beach sand. Of the workable deposits of this type, those in Kerala are important. Total inferred reserves of ilmenite in beach sands have been estimated to be about 103 million tonnes.

18. **Iron Ores :** Iron occurs mainly as haematite (Fe_2O_3) and magnetite (Fe_3O_4). It is exported to Japan, India is one of the few countries in the world which have been endowed with very vast reserves of iron ore. Large deposits of haematite ores are known in Bihar, Orissa, Madhya Pradesh, Mysore and Goa. Magnetite occurs in Tamil Nadu, Mysore, Bihar, Orissa and Andhra Pradesh. Spathic (carbonate) ore is found in West Bengal. The total inferred reserves of all types of iron ores are placed at 21478 million tonnes.

19. **Kyanite :** It contains 56 to 62% Al_2O_3. It is exported to U.K., Denmark, Italy and U.S.A. It is mainly found in Bihar. It is mainly used in the manufacture of refractory bricks for glass and ceramic industry. It is also used in the manufacture of gas burner tips, spark plugs, etc.

20. **Limestone :** Deposits of lime stone of different grades are known in almost all the states. However, important states are Orissa, Bihar, Madhya Pradesh, Madras and Andhra Pradesh. According to the Geological survey of India, the reserves of chemical and cement grade limestone deposits estimated so far are placed at 6,188 million tonnes and 39,992 million tonnes respectively.

    In the flux grade category, 1,111 million tonnes of reserves are reported to be estimated. It is used in the manufacture of cement, chemicals, paper, leather, tanning, glass and other industries.

21. **Magnesite :** It is the principal source of magnesium in the form of carbonates, *i.e.,* $MgCO_3$. Occurrences of magnesite have been reported from a number of places in Tamil Nadu, Uttar Pradesh and Mysore. The total reserves have been estimated at 44 million tonnes in Tamil Nadu, 26 million tonnes in Uttar Pradesh. It is used for the manufacture of refractory bricks and linings. It is exported to U. K., U.S.A., Japan and West Germany.

22. **Manganese ore :** India is very rich in manganese ore. According to the Geological Survey of India, a total of about 202 million tonnes is estimated in Madhya Pradesh, Maharashtra, Mysore, Bihar, Gujarat and Rajasthan. Besides, 123 million tonnes of manganese ore has been estimated in Goa.

It is used in steel, ferromanganese and dry batteries. It is exported mainly to U.S.A., U.K. and Japan.

23. **Mica :** Owing to its special electrical and physical properties, it has a world wide market. Previously, India was the chief supplier of mica to the world, though in recent years several competitors such as Brazil, Tanzania, etc. have come up.

    It occurs in three mica belts covering about 3800 sq km. in Bihar, 3110 sq km; in Rajasthan and 1560 sq. km. in Andhra Pradesh. The best quality of ruby mica comes from Bihar while Andhra Pradesh produces green mica.

24. **Molybdenite (MoS_2) :** It is an important ore of molybdenum. It is found in Chotta Nagpur, Rajasthan, South Travancore, Mysore, Bengal, Trichnapoly and Madras. However, it is imported.

25. **Monazue :** It is the costly thorium ore which occurs in Travancore and India has sole monopoly for it. In addition to thorium, monazite contains ceria and other rare earths. The most important use of thorium is in the production of atomic energy.

26. **Pitch Blende :** It is principal ore of uranium. Although India is poor in this mineral, yet small quantities are found in Bihar and Rajasthan. Uranium is used for manufacture of atom bomb.

27. **Rare Minerals :** Investigations carried out by the Atomic Energy Department showed that the deposit of radioactive minerals in India has over 3.3 million tonnes of ore containing 3,000,000 tonnes of thorium of 10% concentration and 10,000 tonnes of uranium of 0.03 to 0.04 percent concentration. The deposits also contain about 80 million tonnes of ilmenite. Ilmenite is recovered from the beach sands of Kerala State contributing nearly 93 percent of the total production and the Manavala Kumchi are in Tamil Nadu. It is mainly exported to U.S.A. and U. K.

    Beach Sands of Kerala-and Tamil Nadu also contain a very useful radioactive mineral monazite. It contains 8 to 10.5 percent of thoria, 20 to 31% of cerium, 1 to 1.5% of ittlium oil and 21 to 30% of rare oxides. Thorium oxide also forms an important source of atomic fuel. The total reserve of combined rare earth oxide in beach sands has been estimated at 4,00,000 tonnes.

Uranium bearing ores have also been discovered in Himalayan region of Himachal Pradesh and Uttar Pradesh and are being presently assessed. Uranium mine at Taduguda in Singhbhum district treats about 1,000 tonnes ore per day worked by Uranium Corporation of India, a Govt. of India Undertaking, India's total reserves of uranium are currently reckoned at 15,000 tons and of thorium at 5,00,000 tons. The reserves of uranium in India can support about 5-10 million kws. of generating capacity over the life-time of the electric power stations.

28. **Rutile :** It is a mineral containing titanium and is found in Kerala.

29. **Sillimanite or Fibrolite :** It is one of the three crystalline forms of aluminium silicate, Al_2SiO_5 found in Assam. India is very rich in sillimanite and has world wide market. It is used as refractory bricks used in building furnaces, in cement, ceramics, glass, and metal smelting factories.

30. **Steatite :** It is found in Bihar, M.P., A. P., Mysore. It is used in the manufacture of pot stone, in building, in radio, radar television and in ornamental work. It is also used in the manufacture of talcum powder.

31. **Sulphur :** India is very poor in sulphur. India is importing it. However, it is found in the form of pyrites in Sahabad, Bihar and Mysore and Rajasthan. It is used mainly as a raw material in the manufacture of sulphuric acid.

32. **Wolfram :** It is the principal ore of tungsten. It is fauna in Jodhpur (Rajasthan) and Bankura (West Bengal). It is used in the electrical industry and in steel alloys.

33. **Zinc-Blende :** The Zawar mines in Udaipur district of Rajasthan are the only producing sources of zinc ores in the country. India is poor in this mineral and has to depend upon the import.

SOLID FUELS

Solid fuels may be classified as being either natural or artificial. The natural fuels are wood, coal, and other vegetable matter whereas the artificial fuels include coke and charcoal which are essentially allotropic forms of carbon. Let us discuss these solid fuels one by one :

(a) **Wood :** Wood consists of cellulose $(C_6H_{10}O_5)n$, lignite, resins, inorganic materials and water. Water content of wood varies from 25 to 50% depending upon its condition whether dry or freshly cut. Heating of a moist wood is difficult and gives out lesser energy. Thus, the heating value of wood is directly proportional to its water content. Wood bums with long non-smoky flame. Its calorific value varies per kilogram of air dried wood. Wood also possesses little ash content (0.3 to 0.6%). As wood possesses low calorific value, it is very rarely used in industry.

Wood is undoubtedly the first fuel used by man, and is still used in large quantities especially in India, although in recent years its use as fuel in industry has been continuously decreasing. It is cheap and abundant especially in the tropical regions. The chief advantages of wood as fuel are the ready inflammability, low ash content and fairly good flames, the only condition is that the wood should be dry. On burning, wood gives carbon dioxide and water

$$C_6H_{10}O_5 + 6O_2 \rightarrow 6CO_2 + 5H_2O.$$

The flame of burning wood may be coloured by impregnating the wood with nitrates of strontium (red flame), copper (green flame), calcium (brick-red flame) potassium (pale-violet flame), etc. The calorific value of air dried wood ranges from 3000 to 4000 cals/kg.

The main drawbacks for using wood are its lower calorific value and the high cost of transporting at large distances by rail.

(b) **Baggase :** Baggase (crushed cane) and pulp from sugar beets are generally used as fuel in sugar factories for steam raising in boilers. In India baggase is used in almost all the sugar factories.

(c) **Coal :** It is a combustible solid. It is formed by the burial of partially decomposed vegetation in past geological ages. After burial, pressure and temperature change them to coal. Due to the difference in the composition of various plant components different types of coal such as banded, splint, channel and bog-head coal are formed under the earth.

Coal is by far the most important solid fuel. Coal varies greatly in composition and properties; the most important types are peat, lignite, bituminous (soft) coal, and anthracite, or hard coal. Following table gives approximate compositions of the various varieties.

Vast peat logs occur in the temperate humid climates of the world, formed by me accumulation and partial decomposition of vegetable matter under marshy conditions. It is mainly used as a fuel but only where there is a deficiency of coal. Its calorific value changes from 4000 to 5000 cals/kg.

Table 1 : Composition of solid fuels, percentage.

Type of coal	*Carbon*	*Hydrogen*	*Oxygen*	*Ash*
Wood	48.5	6.0	43.5	1.5
Peat	58.0	6.3	30.8	4.0
Lignite	68.0	5.1	19.5	6.3
Soft coal	82.0	5.0	6.0	7.0
Hard coal	90.0	2.5	2.5	4.0

The following are the important types of coal:

(i) **Black Lignite :** It is black in colour and has dull waxy lustre, it ignites easily and has calorific value between 16.8-25.2kJ/kg.

(ii) **Bituminous Coal :** It is so called as it burns with smoky yellow flame like bitumen and its distillation product is coal for which is of bituminous nature. It is black in colour. The calorific value of this variety of coal is 18-20 kJ/kg. It is generally used for steam-raising and by-products manufacturing.

(iii) **Anthracite Coal :** It is the final stage of coal formation from vegetable matter. Anthracite is black and has got lustre. On burning it does not emit any smoke. The calorific value of anthracite is between 19-20 kJ/kg. It is very useful for metallurgy, naval purposes and for slow combustion stoves.

(iv) **Charcoal :** It is produced by carbonisation of wood. In India it is produced by the partial combustion of wood which is generally covered with earth, etc. during combustion. Its calorific value varies between 25 to 33 kJ/kg.

Charcoal was previously used for metallurgical purposes but now it is only used for household purposes.

(v) **Coke :** Coke is formed by the destructive distillation of coal. The calorific value of coke containing 90% of carbon and 0.5% of hydrogen is about 30 to 33 U/kg.

This is generally used for metallurgical processes.

All the forms of coal as well as coke and charcoal burn to give off carbon monoxide, and carbon dioxide.

$$2C + O_2 \rightarrow 2CO \text{ (excess of carbon)}$$

$$C + O_2 \rightarrow CO_2 \text{ (excess of oxygen).}$$

GASEOUS FUELS

Gaseous fuels are considered to be the best fuel partly because of the possibility, of utilizing the heat of the combustion products formed for pre-heating the fuel and the air mixed with it, thus increasing the temperature developed during combustion still more. The increased use of gas for heating purposes both for the household and industrial purposes has made the demand larger than ever.

Gaseous fuels possess all the advantages of liquid fuels. However, some of the main advantages are summarised below :

(i) These can flow through pipes.

(ii) These can be lighted at a moment's notice.

(iii) On burning, they leave no ash.

(iv) They possess high heat contents,

Due to above mentioned reasons, fuel gases are widely used for cooking and heating homes as well as in industry. The most important types of gaseous fuels are :

Producer Gas : It is the cheapest fuel gas though its calorific value is lowest of all the gaseous fuels. It is essentially a mixture of carbon monoxide and nitrogen :

CO	30%
N_2	60%
H_2	2 to 5%
CH_4, CO_2	Rest percentage

Producer gas is prepared by passing air through red hot coke.

The plant used for the manufacture of producer gas is known as producer, it consists of a large air-tight cylindrical furnace (stove),

2-4 m internal diameter and 3-5 m high, lined with fire bricks with an outer mild sheet iron casting.

An air inlet is provided at the base and an exit for the gases near the top. Arrangements for admitting fuel and for removing ash, without admitting air are also provided.

Coal is burnt in this furnace while air enters from the bottom through an inlet as shown in Fig. 2.1. The following reactions take place in the furnace :

(i) Carbon (of the coal) combines with oxygen (of the air) in the lower part of the furnace to produce CO_2.

$$C + C_2 \rightarrow CO_2 + 406KJ \quad ...(1)$$

(ii) Carbon dioxide so formed rises up through upper layers of the red hot coal and gets reduced to carbon monoxide.

$$CO_2 + C \rightarrow 2CO - 163 \text{ kJ} \quad ...(2)$$

(iii) Nitrogen remains unaffected.

The producer gas which is a mixture of carbon monoxide and nitrogen is drawn out at the outer at the upper end of the producer. Combining the two equations (i) and (ii), we have

$$2C + O_2 \rightarrow 2CO + 243 \text{ kJ}$$

or

$$C + 1/2O_2 \rightarrow CO + 121.5 \text{ kJ.}$$

This means that when producer gas is obtained, it still has 121.5 kJ. of heat which can take part for heating purposes. So the process can go uninterrupted if the fuel is periodically removed.

Some CO_2 in the producer gas is invariably present, therefore, the temperature of the fuel has to be kept high 1200°C to effect complete reduction of CO_2.

COMPOSITION, PROPERTIES AND USES OF PRODUCER GAS

Producer gas consisted of mainly carbon monoxide (about 30%) and nitrogen (about 60%) with small amounts of other gases like methane, CO_2 etc. It is poisonous in nature, heavier than air and insoluble in water. It is a non-supporter of combustion, though combustible itself, burning in air to produce CO_2.

The producer gas is used as a fuel where the gas and air are burnt to give heat.

$$2CO_2 + O \rightarrow 2CO_2 + 569 \text{ kJ}$$

Nitrogen acts as a diluent and, therefore, the calorific value of the gas is low.

The producer gas is a poor fuel, but is used largely because of its cheap production.

Water Gas : Water gas is a mixture of carbon monoxide and hydrogen. A small amount of carbon dioxide is also present in mixture. It is obtained by blowing steam through a bed of coke heated to 1050°C.

$$C + H_2O \rightarrow CO + H_2 - 121.5 \text{ kJ}$$

The reaction between coke and steam is endothermic. The red hot coke, therefore, cools down after a few minutes and no more of water gas can be obtained. With the fall of temperature, the reactions take a new turn resulting in the accumulation of CO_2.

$$C + 2H_2O \rightarrow CO_2 + 2H_2 - 71 \text{ kJ}$$

But at 1050°C the percentage of CO_2 is very low *i.e.,* 0.6 to 1.5%. The difficulty is overcome as under:

In modern gas plants, the carbon in the producer is raised to hot bright redness by a blast of air (Air blow) continued for 2 minutes. This raises temperature above 1050°C. This is followed by a jet of steam (steam blow) until the temperature falls to dull redness, continued for 4 minutes. When the air blow is in progress, the producer is not making water gas and the gases are allowed to escape. Thus the manufacture of water gas is intermittent.

The plant used for the production of water gas is a cylindrical vessel made of steel 4-5 m in height, 2-3 m in diameter. The coke is introduced through an opening at the top. There are separate inlets for blowing steam and air alternately from below. The water gas produced is taken out from the exist near the top of the vessel.

COMPOSITION AND USES OF WATER GAS

Water gas consists of mainly carbon monoxide (about 44.2%) and hydrogen (about 48.6%) along with small amounts of other gases such as CO_2 (4.2%) methane (0.8%) and nitrogen (3%).

Water gas has been put to the following uses :

(a) It is used as a fuel gas of high calorific value. The carbon monoxide and hydrogen present in it burn to produce large amount of heat.

$$2CO + O_2 \rightarrow 2CO_2 + 569 \text{ kJ}$$

$$H_2 + 1/2O_2 \rightarrow H_2O + 285 \text{ kJ}$$

(b) Large quantities of water gas used for the manufacture of methyl alcohol in which a mixture of water gas and H_2 when passed over heated zinc oxide as catalyst under a pressure of 200-300 atmospheres produces- methyl alcohol.

$$CO + 2H_2 \xrightarrow{ZnO} CH_3OH$$

(c) A mixture of water gas and producer gas is used for the manufacture of ammonia by Haber's process.

(d) It is also used for the preparation of carburetted water gas.

Semi-Water Gas : Water gas possesses much lower calorific value and hence this has been substituted by semi-water gas.

It is however obtained by passing mixture of steam and air continuously through white hot coke. Thus, the semi-water gas is essentially a mixture of water gas and producer gas.

Uses It is used in steel industry as well as in internal combustion engines.

Carburetted Water Gas : Pure water gas bums with a non-luminous flame. Therefore, it cannot be used for lighting purpose. In order to obtain a luminous flame, the water gas is mixed with hydrocarbon vapours which do not condense on cooling. Gas which is rich in carbon is known as carburetted water gas and burns with a luminous flame.

The plant for the manufacture of carburetted water gas.

(i) **Water gas generator :** Water gas is obtained as above by passing steam over red hot coke at 1050°C in the first chamber. To begin with, the blast of air is blown through this generator (air blow) to raise the temperature to 1050°C. The resulting hot gases (mostly hot producer gas and air) are led into the next two towers, carburetter and the superheater and make them red hot.

(ii) **Carburetter :** The water gas is passed through the carburetter which is a big tower packed with checker brick work in which mineral oil is sprayed from top. The bricks, being hot from the previous operation, make oil vapour and water gas to get volatilized. The mixture is then introduced in the next part of the plant.

(iii) **Superheater :** This is another tower similar to the carburetter packed with brick checker-work heated to redness by the passage of a hot mixture of producer gas and of the air blow. The water gas containing mineral oil vapours is introduced in the superheater. The oil here gets cracked due to the high temperature into simple gaseous hydrocarbons such as ethylene, etc., which bum with a luminous flame.

(iv) **Purifier :** The carburetted water gas is then passed through the purifier which contains hydrated ferric oxide. This removes the H_2S present as impurity in the carburetted water gas (which originates from the sulphur present in coke).

Finally, the gas is scrubbed and collected in gas holders.

Composition and Uses : Carbureted water gas contains a large proportion of gaseous hydrocarbons in addition to the usual carbon monoxide and hydrogen. Carburetted water gas is put to following uses.

(i) It is used for illuminating purposes, since it burns with a luminous flame.

(ii) It is having a high calorific value and is used as an important fuel.

(iii) It is also used for welding purpose.

Oil Gas : This is the gaseous fuel used in laboratories. It is a mixture of gaseous hydrocarbons.

It is prepared from kerosene oil by a process of cracking, *i.e.*, decomposing it at a high temperature, whereby gaseous hydrocarbons are produced , *e.g.*,

$$C_{16}H_{34} \rightarrow 6CH_4 + 5C_2H_2$$

The plant used in making oil-gas.

A thin stream of oil is allowed to flow into the cast iron retort, which has previously been heated to redness in the furnace.

The cracking of the liquid hydrocarbons occurs with the production of gaseous hydrocarbons, methane, ethane, etc. The gas is first passed through a scrubbing box, which retains the tarry matter present in it. It is then collected in gas-holders. Before distribution to the laboratories, the gas is properly aerated, *i.e.*, mixed with air.

Coal Gas : When coal is submitted to destructive distillation, it gives the following products :

(a) *Coal gas* which is a mixture of gases that do not condense on cooling. Its average composition is:

H_2	=	40-50%	by volume
CH_4	=	30-35%	"
C_2H_2	=	4%	"
CO	=	5-10%	"
N_2	=	8%	"

(b) A condensable liquid portion which separates into two layers:

(i) An upper aqueous layer, called ammoniacal liquor, which contains free ammonia, $(NH_4)_2S$, NH_4CN, NH_4CNS and $(NH_4)_2CO_3$.

(ii) A lower tarry layer, called coal-tar, which is a complex mixture, consisting principally of aromatic hydrocarbons, phenols, pyridine and other bases.

(c) Coke, which is the non-volatile residue of carbon and ash left behind in the retorts.

A modern plant for the manufacture of coal-gas, using vertical retorts. The retorts are charged with coal and heated to about 1050°C by means of producer gas. The volatile products leave the retorts at the top and pass through hydraulic-main, where a good portion of tar and ammoniacal liquor condense. From here, the gases pass through condensers. Condensate is collected and separates into two layers on standing. The lighter layer is of ammoniacal liquor and the heavier layer is of coal-tar.

The gas next passes up the scrubber or washing tower, where it is washed by a current of water. NH_3 is removed and ammoniacal liquor collects at the base of the scrubber.

Natural Gas : Methane is the principal constituent of natural gas which issues from wells mainly situated in the oil producing regions. It is an excellent fuel and can be stored at relatively low cost. It is conducted to distant places through long pipes. A typical natural gas might have the following compositions

Methane	85%
Ethane	8%
Propane	4%
Butane	1.5%
Higher hydrocarbons	1.5%

The gas is carried by a network of pipe lines from the field to the factory or wherever, needed.

Blast Furnace Gas : This gas escapes out from the top of the furnace used for the manufacture of pig iron. Its composition is 60 per cent nitrogen, 25 per cent carbon monoxide, and the rest mainly carbon dioxide with a little hydrogen.

REQUIREMENTS OF A ESSENTIAL FUEL

Requisites of a good fuel are :

(i) **Cost of fuel :** It should be cheap. A small amount should be spent on its supply and its combustion.

(ii) **Supply position :** Supply position with regard to availability in sufficient quantity should be considered. Reliability of regular supply is also taken into consideration.

(iii) **No by-products :** It should not give on burning any undesirable products. It means that during combustion a good fuel should not give any objectionable gases.

(iv) **Calorific value :** A fuel must have a high calorific value, since the amount of heat obtained from a definite weight of the substance depends upon this factor.

(v) **Moderate velocity of combustion :** The highest temperature achieved by combustion of a fuel depends upon the calorific value and the velocity of combustion ; with any carbonaceous fuel, the highest temperature obtained is 2800°C. For a continuous supply of heat, fuel must burn with a moderate velocity.

(vi) **Proper ignition point :** A good fuel must have a moderate ignition point since low ignition temperature of fuels is mostly accompanied by fire hazards, while high ignition point makes the fuel safe for transportation and storage but not favourable for starting a fire.

(vii) **Low contents of non-combustible material :** A good fuel must have low contents of non-combustible material as the non-combustible matter is left behind in the form of ash which will naturally decrease the calorific value of the fuel.

(viii) **Highest pyrometric effect :** For an industrial fuel in addition to its heat value (the maximum amount of heat evolved upon complete combustion of one kilogram of the fuel), the highest temperature obtained with the fuel is also important. This temperature is called the pyrometric burning effect and, besides the calorific power of the fuel, it depends to a considerable extent on the physical state (*i.e.*, solid, liquid or gas) of the fuel.

In case of solid fuel, approximately one and a half times the amount of air needed theoretically must be introduced into the furnace for its complete combustion. The excess of air absorbs part of the heat evolved and thus greatly lowers the pyrometric burning effect. A higher pyrometric effect is obtained when a liquid fuel, such as crude oil is burnt. But only gaseous fuel can be mixed perfectly with air and therefore it bums up completely with almost the theoretical quantity of air and gives the highest pyrometric effect.

CLASSIFICATION OF FUELS

Different types of fuels are :

(A) **Solid Fuels :** These include wood, charcoal, coke, coal, etc. These are derived directly or indirectly from plants.

(B) **Liquid Fuels :** Examples of this are petroleum and alcohol. These are more convenient and leave no solid ash.

(C) **Gaseous Fuels :** Combustible gases are regarded as excellent fuels. These are most convenient in use.

Let us discuss these one by one.

FUELS

Any source of heat energy is known as fuel. The term fuel includes all combustible substances obtainable in bulk which may be burnt by means of atmospheric air in such a manner as to render the heat evolved capable of being economically applied to domestic and industrial purposes.

At the same time, it is required that a fuel on burning should not produce excessively objectionable by-products. Hence this discussion is followed by the simple definition of a fuel:

"Any combustible substance which may be burnt to supply heat energy without the products of exclusively objectionable by-products is called a fuel."

The value of a fuel depends upon the quantity and intensity of the heat obtained per unit of fuel. The heat value can be determined by burning a known quantity of the fuel in a calorimeter. Substances used as fuel may be either solid, liquid, or gaseous, and most of the latter contain more or less carbon dioxide.

Heat evolved by burning of fuels can be employed for heating, preparing food and for many industrial processes. However, most of the obtained heat is converted into other kinds of energy and is used to do mechanical work.

Moreover, to put various machines into motion we use, besides thermal power, the power of falling water (white coal) and wind power; there have also been efforts to make direct use of solar energy.

LIQUID FUELS

Although the extensive use of liquid hydrocarbons for fuel is of very recent origin, the use of liquid fuels dates back to very ancient times, when vegetable and animal oils were burnt for the production of light and to some extent for heat. These oils are of little importance for these purposes today because they have been largely supplanted by petroleum products.

Liquid fuels are said to have the following advantages over solid fuels :

(i) Easy to handle

(ii) No ash on combustion

(iii) No danger of any spontaneous combustion.

(iv) Temperature can be increased or decreased by changing the state of burning.

Liquid fuels have disadvantage of high inflammability. Some examples of liquid fuels are :

(i) **Petroleum Fuels :** With the advent of the internal combustion engine, it has been used widely as a rule since 19th century. In the engine, petrol burns in the air and thus liberating heat energy. Besides its use in internal combustion engine in locomotives and steamship; crude petroleum is also used as a fuel in manufacturing plants, and in oil burning furnaces in homes. The complete combustion takes place as follows.

$$2C_nH_{2n+2} + (3n + 1)O_2 \rightarrow 2nCO_2 + (2n + 2)\ H_2O$$

(ii) **Kerosene :** It is one of the products of natural petroleum. Kerosene oil contains paraffins, naphthalenes and aromatic compounds. The number of hydrocarbons ranges from 10 to 16 carbon atoms per molecule. It is used in cooking stoves, kerosene oil lamps and as a fuel for diesel engines and for jet and turbo-jet engines. Its main advantage is that it is very cheap. However, It is not suitable for automobile engines, due to its low volatility flash point which varies from 24 to 66°C although it can be used at an appreciable sacrifice of efficiency.

(iii) **Gas Oil :** It is the fraction obtained from petroleum distillation between kerosene oil and lubricating oils. Boiling range varies between 180 to 450°C. Gas oil contains 17 carbon atom compounds. It is used as an industrial and domestic fuel for heating purposes and for the synthesis of gasoline by several processes.

(iv) **Gasoline :** It is a mixture of hydrocarbons boiling between 38°C and 204°C. Gasoline is obtained from crude petroleum by fractional distillation or by cracking of heavier fractions of crude petroleum or by synthetic processes such as polymerisation of gaseous olefins or alkylation of isobutane with low boiling olefinic hydrocarbons. Gasoline is mixture of petroleum hydrocarbons containing 6 to 8 carbon atoms per molecule. It is mainly used as fuel for internal combustion engines having a spark ignition (automobile and aeroplane engines).

Alkylate or aviation gasoline is used for aeroplane engines. This consists of mainly of iso-octane. It is manufactured by taking two compounds of lower molecular weights than those in gasoline and allow them to react in the presence of cone. sulphuric acid or HP.

$$CH_2{=}\underset{\displaystyle CH_2}{\overset{|}{C}}{-}CH_3 + CH_3{-}\underset{\displaystyle CH_3}{\overset{|}{CH}}{-}CH_3 \xrightarrow[HF]{H_2SO_4 \text{ or}} CH_3{-}C(CH_3)(CH_3){-}C_2{-}CH(CH_3){-}CH_3$$

Isobutylene Isobutane Iso-octane

The complete combustion of iso-octane takes place in the following manner:

$$2C_8H_{18} + 25O_2 \rightarrow 16CO_2 + 18H_2O$$

(v) **Alcohol Fuel :** This is another important liquid fuel. As it has high cost, it is finding less uses as a fuel. However, it gives clean and hot flame. Two important examples of alcohol fuels are wood or methyl alcohol and denatured alcohol.

Wood alcohol is manufactured by the reaction of carbon monoxide with hydrogen in the presence of a suitable catalyst or by the destructive distillation of wood. When it is burnt, it gives carbon dioxide and water.

$$2CH_3OH + 3O_2 \rightarrow 2CO_2 + 4H_2O$$

Denatured alcohol is simply ethyl alcohol but it is made unfit for drinking and other beverage purposes by adding methyl alcohol. However, on burning it gives carbon dioxide and water.

$$C_2H_5OH + 3O_2 \rightarrow 2CO_2 + 3H_2O$$

It has been demonstrated that a small percentage of alcohol added to gasoline gives a fuel that burns reasonably well in an internal combustion engine. It is expected that sometime in the near future, automobile engines may be designed to use alcohol as a fuel in the place of gasoline.

FURNACES

A furnace is the enclosed space for the burning of fuel. Actually it refers to a set up in which a high temperature is produced either by burning of fuel or by passing electricity. Furnaces are of two types

1. Fuel Fired Furnaces.

2. Electrical Furnaces.

Let us discuss these one by one.

1. Fuel Fired Furnaces

In a fuel-fired furnace, high temperature is produced by burning of fuel in a closed space. Such types of furnaces are divided into several different types :

(a) **Blast Furnaces :** Blast furnaces are steel structures lined inside with a refractory material. They are provided at the top with a charging arrangement and an outlet for the exit of waste gases, and at the base with a row of tyres or inlet pipes for airblast.

A typical furnace is a shaft furnace of characteristic shape, about 30 m high with a hearth of 5-8 m in diameter. The furnace is made of steel plates lined inside with fireclay bricks. The hearth may be lined with carbon blocks for durability or with fire-clay, as the rest of the shaft. The charge is introduced as required through a hopper and double bell system at the top. Air is heated by combustion of waste gases and a blast of this air is introduced under pressure through 8-20 nozzels called tuyeres, served by a bustle pipe which encircles the furnace at about 2.5 m from the base.

The down coming charge meets the upward moving hot air blast. Addition of oxygen to the air blast and the application of higher gas pressures throughout to slow down the blast gives increased reaction time, an improved yield and more economical use of fuel. From the furnace, the products are coming out at three places. Molten metal is tapped from the base of the hearth every six hours or so, the less dense molten slag floating on it is tapped more frequently through the separate outlet just above the molten metal and the waste gases, collected at the top of the furnace. The waste gases are purified prior to combustion.

The temperature range in the furnace drops from approximately 1500°C at the tuyeres to 200-300°C at the top. Blast furnaces are frequently used for the extraction of iron and copper from their ores. The chemical changes that occur during extraction of iron are summarised in the labelled diagram of the blast furnace. It is important to note that the slag ($CaSiO_3$) plays an important role by covering the molten iron and so protecting it from being reoxidized as it drips through the highly heated air in the lower part of the furnace.

(b) **Reverberatory Furnace :** In this type of furnace the material is heated to a high temperature not by direct contact with the fuel, but by the hot gases that are produced by the burning of the fuel. Air may be admitted at will, so that the charge can be heated in either a reducing or an oxidizing atmosphere. A typical modem reverberatory smelting furnace is a large, low brick structure, strengthened externally by steel buckstays and tie rods and lined with heat resisting bricks such as magnesite or silica.

The interior forms a single chamber. 30-40 m long, 8-10 m wide and 3-4 m high, which has a roof of bricks commonly held in place by cast iron or steel hangers suspended from the framework overhead. At the end of the furnace, several burners are so placed as to shoot a long intensity hot flame down the chamber between the charge and the roof. The charge is melted mainly by radiation of heat from the gases. Hot gases after imparting their sufficient heat to the charge are dropped in temperature from 1550°C to 1150°C.

They are now passed through a waste heat boiler to provide steam for power or heating purposes. Hot air for combustion is usually supplied by an air preheater, which takes more heat from the products of combustion after the boiler has called them to about 175° in the air preheater.

The air supply and the fuel may both enter the furnace through a common nozzle, or the fuel may be dropped just in front of the burner from a hopper overhead. The furnace operates in compaigns of many months each. In the above mentioned size of the furnace nearly 500 to 1000 tons of charge may be smelted in a single day.

Charging is done every few hours. In reverberatory furnace since the fuel does not come in direct contact with the charge, it can be used for reduction as well as oxidation processes. For reduction the material is mixed with a reducing agent like coke and heated while for oxidation it is heated in a current of air. Thus calcination and roasting are usually carried out in a reverberatory furnace.

(c) **Muffle Furnace :** A muffle furnace is a chamber made of refractory material and surrounded by flames and hot gases all around. The muffle furnace is so designed that flames have no access to the reacting substances. The latter are placed in closed retorts or crucibles and all heating is done from the out side.

(d) **Regenerative Furnaces :** In this type of furnace, the heat carried away by the flue gases is not allowed to be wasted as such. Hot blue gases are passed through a brick work chamber which gets heated and the going out of this chamber are at a much lower temperature.

After some times fuel gases are allowed to pass through hot chamber and the flue gases are made to heat another brick work chamber. The fuel gases preheated in the hot chamber burn and supply more heat. After a certain time, the currents are again reversed.

2. Electrical Furnaces

The term 'electric furnace' is probably the most general term which is used *for a group of heating apparatus using electric energy as a source.* Sometimes the term electric furnace is also used to *cover electric heating equipment/or dielectric heating non-metallic materials, induction heating, infra-red heating, or simple resistance heating of various kinds.*

In such furnaces electrical energy is converted into heat energy ; the heat thus generated is proportional to the resistance encountered by the electric current and to the square of the current passed. These furnaces are largely used where a cheap power is available, and very high temperatures are required, and also for electrolytic reductions.

Electric furnaces are commonly classified as resistance, and arc induction types. The resistance medium for the resistance furnace is usually a solid wire, ribbon or rod or sometimes a liquid electrolyte; the resistance medium for the arc furnace is a gas, and for the induction furnace the medium is generally a solid or liquid.

In resistance furnaces electric current usually does not flow through the material to be heated (*i.e.*, the charge), although in certain electrolytic processes, such as those used in the production of aluminium, the current does flow through the electrolyte. Electric current may flow through the charge in certain arc furnaces, or the current flow may be confined to the gas phase above the charge. Finally, in induction heating electric current generally passes through the charge, except in the case of some melting furnaces in which current flows through the crucible and the charge is heated indirectly.

It is important to note in the study of furnaces that heating usually takes place by a subtle interplay of the three modes of hat transfer-convection, conduction and radiation.

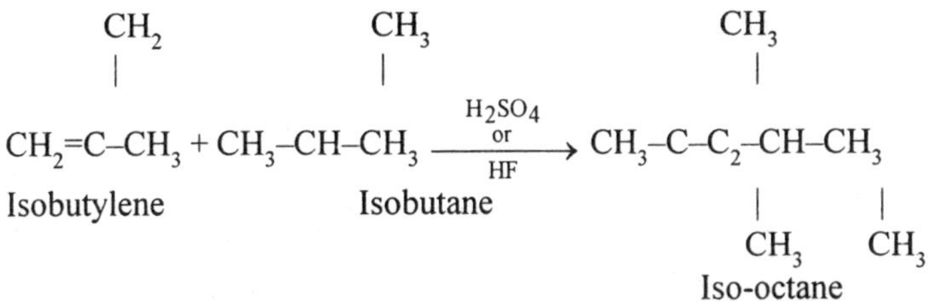

Classification : There are various ways of classifying electric furnaces. The most convenient way of classification is found in the manner in which electric energy is converted into heat. According to this type, there are three main types :

(a) **Arc Furnaces :** In arc furnaces heat is generated in electric arc, which by its nature can bum only in gaseous atmosphere. In this type, a temperature of over 3000°C can be attained. Arc furnaces are of two types :

In the direct-heat arc furnace, arc is used to heat a gas in which the arc is burning and in the indirect heat furnace, the arc burns above the charge. Thus, in the indirect type, the arc radiates heat towards the charge by conduction, convection and or radiation. In simple words, in the direct-heat furnaces the charge forms a part of the electric-circuit while in the indirect-heat furnaces, the charge does not form a part of the electrical heating circuit.

Applications : Direct-arc furnaces are used for the production of nitrogen. It is also used primarily for melting steel. It is also used for production of ferro-alloys of alumina and phosphorus and in other electrochemical processes.

Indirect-heat arc furnaces are used for some chemical processes and for smelting of metals.

(b) **Resistance Furnaces :** In this type of furnaces, heat is generated by the resistance in the electric circuit. In some cases the material forming the charge may act as the resistance and does self-heating. In some of the furnaces, the body of the furnace is made up of a resistance material and this material may cause heating. Resistance furnaces are of two types :

Direct-heat resistance furnace : The direct type comprises of all furnaces in which heat is generated directly in the material to be heated. Direct-heat resistance furnaces can be grouped into,

(i) Induction furnaces,

(ii) Electrode furnaces.

The current is transmitted through the charge without physical contact between the latter, (*i.e.,* charge) and electric supply system. In electrode resistance furnaces, the current is applied to the charge by means of electrodes. In this type of furnaces, the electrodes need not touch the charge when they are operating at high frequency. In other types when they are operating at normal frequency, physical contact between electrode and charge is necessary.

Indirect heat resistance furnace :: In the indirect-heat resistance furnaces, the charge is not part of heating circuit. Such furnaces are also called *resistor* furnaces.

Applications : Direct-heat resistance furnaces are used for melting and heating of metals and for the brazing, soldering and high-frequency heating of electrically non-conducting materials. Typical applications are the manufacture of graphite, carbon and the various carbides like silicon carbide boron carbide, etc.

Another broad field of application of direct resistance heating is the melting of glass. Molten glass is a good conductor of electricity and thus can serve as resistor.

According to the main mode of heat transfer, one may distinguish three types of indirect heat resistance furnaces. These are :

1. Radiation-Type.
2. Convection-Type.
3. Conduction-Type.

Applications for the three types of indirect-heat resistance furnaces are fairly different. For example,

1. Radiation type furnaces are used for drying of paints, enamels, etc. on surfaces and are also used in the heat treatment of high-alloy steels for forging and brazing and for sintering powdered metal.
2. The use of convection-type furnaces is constantly growing. These are mainly used in drawing and hardening steel, age hardening and other heat-treating operations, for aluminium and light metals.
3. Conduction furnaces are mainly used for annealing of copper and brass, annealing of aluminium and light alloys, and annealing of silver and bracing of steel and some other alloys.

(c) **Electron beam furnace :** In this type, heating by electron beams is carried out by causing a cathode to emit electrons which through suitable guidance (electric and magnetic fields) are directed to impinge on the charge.

Applications : It is used for melting or localised heating. Its relative high cost limits its applications for melting to refractory metals requiring an intent atmosphere. This technique is new and may be used in the new fields.

(d) **Combination type furnace :** A commercial furnace may be a combination of the several above mentioned types, out of which the most common is a combination of an indirect-heat arc furnace and direct-heat resistance furnace.

2

Lanthanoid Contraction

INTRODUCTION

In this series the size of La is maximum and that of Lu minimum. The decrease in size, though continuous, is not regular. The reason of lanthanide contraction can be traced to imperfect shielding of one 4*f*-electrons increases by another in the same sub-shell. Table 1 gives the atomic and ionic radii of lanthanides. In moving along the lanthanide series from Ce(58) to Lu 71 a regular decrease in the size of the atom/ion with increase in atomic number is observed. This decrease in size is called the *lanthanide contraction.*

On moving along the lanthanide series the number of 4*f*- electrons increases by one unit at each step and the imperfect shielding Z_{eff} increases causing a contraction in electron cloud of the 4*f*-subshell. Ionic radii change from 1.06 Å to 0.85 Å. Thus, lanthanide contraction plays a significant role in chemistry of lanthanides. The important consequences of it are :

(i) There is a steady decrease in the ionic size.

(ii) There is a slight increase in electronegativity of the trivalent ions.

(iii) The E° values for $M^{3+} + 3e \rightarrow M_{(g)}$ increases regularly from lanthanum – 2.52 V to 2.25 V for lutetium.

From the figure we find that a general decrease in size (except Eu and Yb) takes place. This can be explained by the fact that lanthanide atoms contribute three electrons towards the formation of metallic bond whereas Eu and Yb contribute only two electrons towards bond formation.

Table 1 : Atomic and Ionic Radii of Lanthanides

Elements	*At. radii (M) (Å)*	*Ionic radii (M^{3+}) (Å)*
La	1.63	1.06
Ce	1.65	1.03
Pr	1.65	1.01
Nd	1.64	1.00
Pm	–	0.98
Sm	1.66	0.96
Eu	1.65	0.95
Gd	1.61	0.94
Tb	1.59	0.92
Dy	1.59	0.91
Ho	1.58	0.89
Er	1.57	0.88
Tm	1.56	0.87
Yb	1.70	0.86
Lu	1.56	0.85

Effects of Lanthanide Contraction

Lanthanide contraction plays a significant role in the chemistry of lanthanides and heavier transition elements. The important consequences of lanthanide contraction are described as follows :

(a) **Similarity among lanthanides :** These is a very small change in the size of ions in lanthanide series and there is no change in outermost shell so there is a close resemblance in the chemical properties of lanthanides.

(b) **Similarity in properties of Y, Ho, and Dy :** The ionic radii of Dy is 0.91 Å and that of Ho is 0.89Å. The ionic radii of Y is 0.90 Å. Due to this the three elements are very similar in properties. They occur together in minerals and due to a similarity in chemical properties it is very difficult to separate them.

(c) **Close resemblance in the properties of second and third transition series :** Because of *lanthanide contraction* the sizes of the elements that follow the lanthanides are close to those of the previous row. Because of this properties of these elements are very similar. Thus the following pairs have similar properties Zr – Hf, Nb – Ta, Mo – W etc. In other words, *transition elements of the third series (i.e., the transition elements following the lanthanides) have virtually the same atomic and ionic sizes as the corresponding elements just above them in their respective subgroups.*

Thus, zirconium and hafnium are having almost the same size inspite of the fact that hafnium was expected to be much bigger than zirconium (as in the usual case within a group). The usual increase in the size of hafnium gets cancelled due to the intervening lanthanide contraction. With the result hafnium (in IV A) is having nearly the same size as zirconium just above it in the second transition series. The same thing happens to Ta (in VA), W (in VIA), etc. *Because of equality in size of Zr and Hf, Nb and Ta, Mo and W, etc. the two elements of pair are having almost the same properites and hence they are found together in nature and become difficult to separate.*

Table 2 : Atomic (Covalent) radii (in Å) of some elements reflecting the effect of lanthanide contraction on the atomic radii of the third series transition elements

IIIA		*IVA*	*VA*	*VIA*	*VIIA*		*VIII*		*IB*
Sc		Ti	V	Cr	Mn	Fe	Co	Ni	Cu
1.64		1.45	1.31	1.25	1.37	1.24	1.25	1.25	1.28
Y		Zr	Nb	Mo	Tc	Ru	Rh	Pd	Ag
1.80		1.59	1.43	1.36	1.32	1.33	1.35	1.38	1.44
La	**Ce to Lu**	Hf	Ta	W	Re	Os	Ir	Pt	Au
1.88	**1.82 1.73**	1.56	1.43	1.37	1.37	1.34	1.36	1.39	1.44
	Lanthanides								

(d) Basicity differences : A decrease in basic strength of oxides and hydroxides is observed as the atomic number increases. So, $La(OH)_3$ in most basic whereas, $Lu(OH)_3$ is least basic. This is explained on the basis of lanthanide contraction. The successive decrease in size of the lanthanide cations (lanthanide contraction) increases successively the covalent character, (*i.e,*. decreases the ionic character) between the lanthanide ion and the hydroxide ion (Fajan's rules), thereby reducing the basic character of the lanthanide hydroxides, Basicity decreases as the ionic radius decreases.

Thus, basicity of Ln^{3+} ions may be expected to decrease in the order : $La^{3+} > Ce^{3+} > Ce^{3+} > Pr^{3+} > Nd^{3+} > Pm^{3+} < Sm^{3+} > Eu^{3+} > Gd^{3+} > Tb^{3+} > Dy^{3+} > Ho^{3+} > Er^{3+} > Tm^{3+} > Yb^{3+} > Lu^{3+}$.

These differences in basicity are reflected in :

(i) *hydrolysis of ions,*

(ii) *solubility of salts,*

(iii) *thermal decomposition;*

(iv) *formation of complexes.*

(i) **Hydrolysis of ions :** We know that more basic ions hydrolyse with difficulty whereas less basic ions hydrolyse readily. A slight increase in degree of hydrolysis in solution is observed from La^{3+} to Lu^{3+}.

(ii) **Solubility of salts :** The precipitation of hydroxides takes place at progressively decreasing pH values from La^{3+} to Lu^{3+}. The values for precipitation from their nitrate solutions at 25°C are La3+ at pH of 6.82, Gd^{3+} pH of 6.83 and Lu^{3+} at pH of 6.30.

(iii) **Thermal decomposition of oxy salts :** The thermal stability of trinitrate salts decreases as we move along the series.

(iv) **Formation of complexes :** The tendency to form complex ions increases in moving along the series.

(v) **Occurrence of yttrium (a transition element) with heavier lanthanides :** As the size of Y^{3+} ion (~0.90°) has been comparable to the heavier lanthanide ions. viz. Tb^{3+} Dy^{3+}, Ho^{3+} and Er^{3+}, the former is found with the latter in natural minerals. Actually, the crystal structure, solubility nd chemical properties of the yttrium compounds have been found to be so close to those of the corresponding compounds of the heavier lanthanides that the heavier lanthanides are commonly termed as yttrium earths.

SEPARATION OF LANTHANIDES

Similarities between laboratory and technical approaches make it impractical, either here or later, to distinguish between them. Processing amounts to cracking the mineral recovering the lanthanides (and thorium) removing thorium if it is present, and separating the lanthanides. It is ordinarily most convenient to remove cerium before fractionating the other lanthanides.

Cracking and Recovery Procedures

The method is the most highly worked. Nearly every possible anion that forms soluble salts with the tripositive ions has been tried with the hope of developing an efficient crystallisation separation.

The method is based upon the fact that the lanthanides form the isomorphous salts which differ not only in solubilities but also in the fact that decomposed by concentrated sulphuric acid to give the water-soluble sulphates along with the evolution of carbondioxide and hydrogen fluoride gas. Now since monazite is the mineral used for the extraction of the elements, the cracking details of only this mineral will be discussed in the present book.

As described earlier the monazite may be cracked by two different methods.

(a) **By means of sulphuric acid :** Monazite is digested with concentrated sulphuric acid at elevated temperature for several hours to give a mixture of sulphates containing phosphoric acid. The resulting grey mud or paste is treated with cold water when SiO_2, TiO_2 and $ZrSO_4$ remain in the residue. The filtrate containing lanthanides, thorium, phosphate, sulphate, etc. ions is diluted to the proper acidity to precipitate thorium as its phosphate. Lastly, the lanthanide ions are freed from phosphate ions by precipitating the former as hydrous hydroxides.

(b) **By means of sodium hydroxide :** Monazite sand is heated at about 140° with 73% sodium hydroxide. The mud so obtained is diluted with water and then cooled to about 100°C when the lanthanides are precipitated s hydroxides along with SiO_2, TiO_2 and $ZrSO_4$ etc. The filtrate containing phosphate and silicate ions is discharded. The residue containing $Ln(OH_3)$ is heated with hydrochloric acid when the lanthanides and thorium ions go into solution whereas titanium and zirconium ions go into residue.

Removal of thorium : Thorium can be removed from the solution obtained above by the (a) or (b) method by the following methods :

(a) By diluting solution to the proper acidity which precipirates out thorium as thorium phosphate (as described earlier).

(b) By adding hydrofluoric acid to the strongly acidic solutions having lanthanides and thorium ions. Under such conditions lanthanide fluorides (LnF_4) go into solution while thorium fluoride is precipitated (as described earlier).

(c) By extracting with non-aqueous solvents. For example, the solution of lanthanides and thorium is extracted with tri-n-butyl phosphate or certain higher amines when thorium is more readily transferred to the non-aqueous phase than the lanthanides.

(d) By selective anion exchange : The solution is passed through an amine type exchanger whereby the thorium is removed as a nitrato or sulphato complex ion.

Removal of Cerium

Cerium is generally removed before the actual separation of the lanthanides. Although there are various methods by which the cerium can be separated from the other lanthanides, only the two important ones are given below :

(i) The mixture of lanthanide oxides is extracted with dilute nitric acid to dissolve the more basic sesquioxides, Ln_2O_3, and leave a residue of CeO_2. The crude cerium dioxide so obtained is dissolved in concentrated (85%) nitric acid and then transferred to an excess of dilute sulphuric acid, when the cerium is precipitated as the red, basic nitrate, $Ce(OH)NO_3)_3.3H_2O$.

(ii) The mixture of the lanthanides is extracted with tri-n-butyl phosphate, when Ce^{4+} along with thorium goes into non-aqueous phase, while the other lanthanides^{3+} go into aqueous phase. Furthermore, reduction of Ce4+ by means of sodium nitrite results Ce^{3+} to the aqueous phase, while the thorium again remains in the non-aqueous phase.

Separation of the Lanthanides

Separation may be effected by one or another or some combination of the following general procedures involving :

1. Fractional crystallization of isomorphous salts.
2. Basicity differences.
   (a) Fractional precipitation from solution.
   (b) Fractional thermal decomposition of salts.
   (c) Ion exchange.
   (d) Solvent extraction.
3. Selective oxidation or reduction.
4. Physical differences.

It is important that the path taken by each component should be followed analytically as fractionation proceeds preferably by a rapid procedure that is specific for each cation and does not destroy or alter the sample. Spectrophotometric procedures are the most generally useful. Each absorbing tripositive species has bands free from interference by other cations that adhere to the *Beer-Lambert* relationship.

$$\log_{10} I_0/I = kcl \qquad ...(1)$$

where I_0 and I are intensities of the incident and transmitted light, c is concentration, l is the light path, and k is a constant. Secondary X-ray fluorescence is also useful, but emission spectrographic procedures are best applied only to the detection and determination of small quantities of contaminants in highly purified samples.

Magnetic susceptibility data are of analytical significance only for binary mixtures of paramagnetic substances of known moments or of a paramagnetic and a diamagnetic substance. Average atomic weight, based usually upon the result of analyses of the mixed oxalates for both lanthanide and oxalate content, although of historical importance is of quantitative significance only for a binary mixture or a sample containing yttrium. The lower atomic weight of yttrium (88.905) compared with those of the lanthanides (138.91-174.97) permits one to follow it by this means.

Inasmuch as little of the early work on separations could be followed quantitatively, it is impossible to judge the efficiencies of many of the procedures described.

Fractional Crystallization : The quantity of work reported and the amount of information accumulated relating to this classical procedure are truly overwhelming. It is probable that every anion found to form soluble salts with the tripositive ions has been investigated in the hope of developing an efficient crystallization separation.

Inasmuch *as only those anions yielding truly isomorphous salts that are easy to crystallize, have reasonable and temperature dependent solubilities, and possess measurable solubility differences as the Ln^{3+} ion* changes can be effective, the number of really successful systems must be limited.

Among the more effective compounds are :

1. The double ammonium nitrates, $Ln(NO_3)_3.2NH_4NO_3.4H_2O$ for the removal of lanthanum and the separation of praseodymium from neodymium.
2. The double manganese nitrates, $2Ln(NO_3)_3.3Mn(NO_3)_2.24H_2O$ for the separation f members of the cerium group but not the yttrium group.
3. The bromates, $Ln(BrO_3)_3.9H_2O$ and the ethyl sulfates, $Ln(C_2H_5SO_4)_3.9H_2O$ for the separation of members of the yttrium group.

Crystallization in each instance is from water. The procedure followed with the double ammonium nitrates is typical. The appropriate solution containing the mixed cations is evaporated until about half of the salts present will crystallize upon cooling to room temperature. The crystals are separated from the mother liquor. Each fraction is then recrystallized, the initial crystals by dissolving in water and evaporating until half the material deposits and the mother liquor by evaporating directly. Four fractions result.

The mother liquor form the first crystals in this step is then combined with the second crystals, and the operation repeated. The least soluble double ammonium nitrate (that of La^{3+}) concentrating at the *head* (crystal end) of the series, the moderately soluble ones (Pr^{3+}, Nd^{3+}) at the centre, and the most soluble one (Sm^{3+}) at the *tail* (solution end). It is apparent that a triangular scheme of this type can expand into a large number of fractions of steadily decreasing content of lanthanides unless some means of control is used. In practice, it is found desirable to omit certain portions of the scheme, thus permitting removal of one lanthanide in a state of purity and treatment of remaining mixed species as a single species. It is found that if the same weight fraction of a given component is allowed to crystallize in each step, fractions of repeating total composition result. These may be removed and combined, and more of the original mixture may be added when a fraction that repeats its composition is obtained.

Fractional crystallization is normally most effective at the lanthanum end of the series where solubilities differ the most. Pure lanthanum is quite readily separated as the double ammonium nitrate, but no other fractional crystallization procedure is now of commercial importance. Generally, separations in the samarium-europium-gadolinium region and within the yttrium group and particularly tedious and inefficient; pioneering studies often involved hundreds or even thousands of crystallizations.

If bismuth (III) nitrate is added to a mixture of cerium group nitrates containing magnesium ion, its double magnesium nitrate crystalizes between the isomorphous ones of samarium and europium. From the resulting Sm_{3+} –Bi^{3+} and Bi^{3+} – Eu^{3++} mixtures, bismuth can be removed readily with hydrogen sulfide. Bismuth (III) is apparently unique among the non-lanthanides as a separating ion, but easily removable lanthanides (*e.g.*, Ce^{3+} between La^{3+} and Pr^{3+} in the double ammonium nitrates or Pr^{3+} between Er^{3+} and Tm^{3+} in the bromates) may function similarly.

OCCURRENCE OF LANTHANIDES

The abundance data indicate only that as a group and, in certain instances as individuals. The lanthanides are potentially available in unlimited quantities. A complete picture of availability, however, must include also data on concentrations of these species in economically interesting deposits.

Here, the nature of the minerals available, the ease with which such minerals can be cracked and thus cost of processing, the geographical location of ore bodies, and the markets of products are all of importance.

As may be expected from similarities in crystal radii, oxidation state, and general properties, each known lanthanide mineral contains all members of the series (except, presumably, promethium). This is true of even cerium, or gadolinite, although the historical developments available at present might suggest otherwise. Monazite both because of its availability and substantial thorium content, is the most important of the mineral sources. Although massive monazite crystals are found in many pegmatites, in only a few instances (notable in South Africa) are they present in sufficient quantity to permit vein mining. Natural weathering and gravity concentration processes (Sp.gr. of monazite =5.2) however, have produced sizable placer deposits of monazite sands. The most notable of these are in Travancore, where the beach deposits are seemingly limitless and regenerate with each crest of tide.

FRACTIONAL PRECIPITATION

This is somewhat more limited in practicality than fractional crystallization because of the added tedium of removing and redissolving precipitates. Although fractional precipitation reflects the basicity order more nearly exactly than fractional crystallization, yttrium is much more migratory in behaviour during precipitation than during crystallization.

All separations by fractional precipitation depend fundamentally upon basicity differences. However, these differences can be operative only when equilibration between solution and solid phases is complete. This condition seldom pertains in practice, for addition of a precipitating anion to the mixed lanthanides results in a localized excess of reagent and precipitation of all the species present. Redissolution may then be too slow or too far from complete to permit true equilibration before more reagent is added.

For the specific case of a gaseous reagent such as ammonia, this problem is avoided by introducing the gas extensively diluted with unreactive nitrogen. More generally it is solved by generating the precipitating ion slowly and homogeneously throughout the solution. Cases in point included hydroxyl ion generated by the cathodic decomposition of water; carbonate and hydroxyl ions formed by adding urea and heating; carbonated ion produce by thermally hydrolyzing trichloroacetate ion; and oxalate ion produced by thermally hydrolyzing dimethyl oxalate.

Precipitates so formed are commonly granular and easy to handle. Control of the precipitating reagent by addition of another species that may give a complex species has the same effect. This is exemplified by the use of zinc or cadmium ion with ammonia. Similarly, addition of solid oxides (*e.g.*, MgO, Ln_2O_3) or carbonates (*e.g.*, $PbCO_3$, $ZNCO_3$) provides a slowly increasing and controlled source of hydroxyl ion. True basicity effects are also achieved by adding a weak complexing agent or insufficient complexing agent (*e.g.*, $EDTA^{4-}$ or NTA^{3-}) to ties up all of the lanthanide ions before adding the precipitant. In either case, the least readily complexed cation is the most available for precipitation. Although fractional precipitation was once of particular importance in separating yttrium from the heavier lanthanons, it is now of little practical importance.

Fractional Thermal Decomposition of Salts

The temperature at which an oxy salt (*e.g.*, nitrate, sulfate, acetate) is converted to an oxide or basic salt of; lower solubility decreases with decreasing basicity of the tripositive cation. Fusion of the mixed titrates followed by leaching with water rapidly concentrates yttrium in the more basic fractions and effects where oxidation can occur (*e.g., with* Ce or Pr) but not in most other cases.

Ion Exchange Method

Ion exchange is the reversible interchange of interchange of ions of similar charge furnished from the solution with ions present in a solid matrix of the exchanger (synthetic zeolites or resins) without any other significant chemical changes. In 1910, the German chemist Gans applied it to the softening of water using the mineral zeolite.

He was also able to prepare synthetic zeolites, which showed greater activity. However the aluminosilicate exchange (zeolites) were not able to interchange with hydrogen ions efficiently. In 1935, Adams and Holmes observed the certain *synthetic resins* can perform ion exchange.

These synthetic are organic condensation products of phenol-sulphuric acid formaldehyde and related substances. They are hard, granular substances with wide molecular structure. When phenol and formaldehyde are condensed in the presence of an acid or an alkali, resins are formed. By introducing more acid groups to the phenol, more strongly acidic resins are prepared.

Similarly by condensing polyamines and aldehydes, resins of basic nature are obtained. The resins are also known as *cation ion exchangers* where H^+ is the exchangeable ion. Similarly the basic resins are known as *anion ion exchangers* where OH^- is the exchangeable ion.

The acid characteristic groups of various cation exchangers are sulphonic–SO_3H, carboxylic–COOH and phenolic –OH. The basic characteristic groups of various anion exchangers are amino – NH_2, –NHR,–NR_2 or NR^{3+} thus giving strongly basic character.

THEORY AND MECHANISM

A typical ion exchange resin sulphonated polystyrene, denoted by HR may be considered. When solution containing the Ln^{3+} is passed down the column, the exchange of cations Ln^{3+} with protons of the resin takes place and an equilibrium is established as follows :

The rate of ion exchange is

(i) Directly proportional to the concentration of the ions in the solution,

(ii) Directly proportional to the charge on the ions in the solution, and

(iii) Inversely proportional to the size of the hydrated ion.

Hydration is also inversely proportional to the size. Hence, the degree of hydration increases from La^{3+}, Since the ionic size decreases from La^{3+} to Lu^{3+}. This means that the radius of hydrated La^{3+} is the smallest and that of hydrated Lu^{3+} is the largest among the lanthanides.

As an example, let us consider an equimolar aqueous solution of tripoitive lanthanide ions (Ln^{3+}) passed through a cation exchanger. The concentration and the charge of the ions being identical, the exchange depends only on the size of the ions hydrated ions. The hydrated La^{3+} being smaller is exchanged first and the hydratred Lu^{3+} being the biggest exchanged last.

After the lanthanide ions are adsorbed as different bands on the resin column, they re to be released from the column for their individual separation. This releasing process is known as *elution.* The solution obtained a result of elution is known as eluate. Elution is brought about by the solution containing H^+ ions or any other cations. However, Russel showed that when a mixture of several lanthanides (Ln^{3+}) was adsorbed on a solid exchanger, they could *not* be easily separated by eluting them by the addition of sodium chloride.

The work of Russel and Tompkins has shown that if the mixture of Ln^{3+} was adsorbed as a band at the top of the acid exchange column and subsequently eluted with an appropriate, complexing agent like buffered citric acid solution the separation of Ln^{3+} into individual bands takes place on the column and the individual cations are finally removed from the column in the *reverse order* of their atomic numbers.

Ion Exchange Eaquilibria the Significance of Complexing Agents

In order to understand the separation of individual bands of lanthanides on the solid exchanger, we should know why the addition of a complexing agent of a complexing agent increases the capacity of an exchanger to fractionate (separate) these bands. We have also to bear in mind that the efficiency of the separation is obtained from the magnitude of its separation factor a. For two different lanthanides, Ln and Ln^1 being changed from an initial concentration (1) to a final concentration (2), the separation factor can be defined as follows :

$$a = \frac{C_{Ln(2)}/C_{Ln^1(2)}}{C_{Ln(1)}/C_{Ln^1(1)}} = \frac{C_{Ln(2)}/C_{Ln^1(1)}}{C_{Ln(1)}/C_{Ln^1(2)}} \qquad ...(1)$$

where C_{Ln} is the molar concentration of lanthanide.

If the value of $\alpha = 1$, there is no separation. However, the efficiency of the process is directly proportional to the departure of value of α from unity, *i.e.,* the value of α should be either more or less.

We have seen the equilibrium process established when a solution containing M^{3+} (aq) ions is passed down the solid exchanger. Let us consider specifically the distribution of a lanthanide between a solid exchanger (s) and aqueous phase Ln^{3+} (aq) for an equilibrium process. Thus

$$Ln^{3+}(aq) + 3HR(s) \rightleftharpoons LnR_3(s) + 3H^+(aq)$$

may be written as

$$\lambda_0 = \frac{\text{conc. of} Ln^{3+} \text{in resin}}{\text{conc. of} Ln^{3+} \text{ in aq. medium}} = \frac{C_{Ln^{3+}(s)}}{C_{Ln^{3+}(aq)}} \qquad ...(2)$$

where λ_0 is the distribution coefficient. For the sake of convenience. Let us ignore charges and rewrite the above relation as

$$\lambda_0 = \frac{C_{Ln(s)}}{C_{Ln(aq)}} \qquad ...(3)$$

Similarly for the next lanthanide, we may express as :

$$\lambda'_0 = \frac{C_{Ln^1(s)}}{C_{Ln^1(aq)}} \qquad ...(4)$$

and hence the separation factor is

$$\alpha = \frac{\lambda_1^0}{\lambda_0} = \frac{C_{Ln^1(s)} C_{Ln(aq)}}{C_{Ln(s)} C_{Ln^1(aq)}} \qquad ...(5)$$

under the given conditions, the value of α rarely departs appreciably from unity.

Now let us consider the addition of a *complexing group.* The formation of chelated ions involves additional equilibria in solution in the following stepwise manner.*

$$Ln + AA \rightleftharpoons Ln\ (AA) \qquad ...(5A)$$

$$Ln\ (AA) + AA \rightleftharpoons Ln\ (AA)_2 \qquad ...(5B)$$

$$Ln\ (AA)_2 + (n\text{-}2)\ AA \rightleftharpoons Ln\ (AA)n \qquad ...(5B)$$

By adding (i), (ii) we can have the equilibrium for the final substance as

$$Ln + nAA \rightleftharpoons Ln\ (AA)n \qquad ...(5D)$$

Similarly, $$Ln^1 + nAA \rightleftharpoons Ln^1\ (AA)n \qquad ...(5E)$$

Here AA is the complexing group and in both equations (iv) and (v) the charges of the complex ion have been ignored. The displacement of each such equilibrium and hence the stability of the ion with the respect to its components is calculated quantitatively by the magnitude of the suitable equilibrium constant.

For equation (iv), *Formation* or *stability* constant for the complex $Ln(AA)_n$ is expressed as

$$K_{Ln}(AA)_n = \frac{C_{Ln(AA)n(aq)}}{C_{Ln(aq)} C_{Ln(AA)(aq)}}$$

$$= \frac{\text{concentrtion of the complex formed}}{\text{conc. of Ln ion} \times \text{conc. of complexing agent}} \qquad ...(6)$$

The total concentration of a lanthanide ion in solution is the sum of the concentration of the uncomplexed ion $[C_{Ln(aq)}]$ and the complexed ion, $[C_{Ln(AA)n(aq)}]$. The distribution between the solid ion exchanger, (s) and the solution containing lanthanide ion can be obtained by rewriting equation (3) to include both the uncomplexed and complexed ions which are present in the solution.

$$\lambda = \frac{C_{Ln(s)}}{[C_{Ln(aq)} + C_{Ln(AA)(aq)n}]} \qquad ...(7)$$

Thus, the formation or stability constant for the *complexed* ion, $Ln(AA)_{n(aq)}$ can be expressed by rearranging equation (6), when we get

$$C_{Ln(AA)n(aq)} = K_{Ln(AA)n}\, C_{Ln(aq)} C_{Ln(AA)(aq)} \qquad ...(8)$$

By substituting the value of $C_{Ln(AA)(aq)n}$ from equation (6) in equation (7) we get

$$\lambda = \frac{C_{Ln(s)}}{C_{Ln(aq)} + K_{Ln(AA)_n} C_{Ln(aq)} C^{n}_{AA(aq)}} \qquad ...(9)$$

$$\therefore \lambda = \frac{C_{Ln(s)}}{C_{Ln(aq)}\,[1 + K_{Ln(AA)_n} C^{n}_{AA(aq)}]} \qquad ...(10)$$

But $= [\lambda_0$ from equation (3)]

Introducing λ_0 in equation (10), we get

$$\lambda = .\frac{\lambda_0}{[1 + k_{Ln(AA)_n} C^{n}_{(AA)(aq)}]} \qquad ..(11)$$

In the case of sufficiently large formation or stability constant, KLn(AA)n equation (11) can be simplified as

$$\lambda = .\frac{\lambda_0}{k_{Ln(AA)_n} \cdot C^{n}_{(AA)(aq)}} \qquad ..(12)$$

Here 1 has been neglected from the denominator as the formation or stability constant, $k_{Ln(AA)n}$ is large.

Equation (12) shows the influence of the complex ion on the distribution of ions between the solid resin and the aqueous solution. Similarly, for Ln we can express its formation or stability constant. Thus, we may now define the separation factor (a) as

$$\alpha = \frac{\lambda'}{\lambda} \qquad ....(13)$$

If the complexing agent yields chelate products of different stabilities with two or more lanthanide ions their separation is markedly improved due to more efficient elution. When a single complex product is formed as a result of reaction of the lanthanide ion with the complexing agent in 1: 1 mole ratio, the above simple considerations are applicable. This is no doubt true of the amine polycarboxylates, but it is not so in the case of the citrates since so many different compositions are possible depending upon pH. However, in this discussion it is presumed that (i) the complex products are sufficiently stable so that the concentration of an uncomplexed metal ion in solution is negligible as compared to that of the complex product and (ii) the pH is appreciably high and thus the equilibria of the following type can be neglected.

$$Ln(AA)_n^{m-} + H^+ \rightleftharpoons HLn(AA)_n^{(m-1)-}$$

Thus, the elution f the adsorbed ions consists of the successive steps of displacement from the resin in the form of the complex product and the readsorption from the complex product further down the exchanger column. Since both the equilibria are influenced by the stability of the complex products, separation of lanthanide ions into individual bands finally takes place.

A remarkable improvement takes place if the eluting solution also contains a complexing agent such as EDTA which can form complexes of varying stability with lanthanide ions. (The complexing agent is 0.1 to 0.2 M ethylene diamine tetraacetic acid which has been adjusted to pH 8 with ammonia). When the elution solution first percolates through the lanthanide ions pass into solution and are replaced in the resin by NH^+_4 from the complexing agent. When the solution runs through the lower edge of the band, the displaced lanthanide ions pass through first, and exchange with hydrogen ions on the resin below; but are then redissolved by the fresh NH^+_4 (aq) following up, so the whole band is gradually carried down the release lutetium ions.

OXIDATION POTENTIALAND OXIDATION STATES

The ease of formation of the various oxidation states in solution can be revealed by the values of the standard electrode potential, E°. These values (in volts) for different couples of lanthanides such as Ln^0 + → Ln^{3+} +3e^-, Ln^{2+}→ Ln^{3+} e^- and Ln^{3+} → Ln^{4+} + e^-, for 1M per-chloric acid at 25°C have been included in Table 1.3. Here Ln represents the elemental lanthanide. Ln^{2+}. Ln^{3+} and Ln^{4+} refer to the di-, tri-and tetrapositive lanthanide ions. From this table, we observe that :

(i) The high positive values of the oxidation electrode potentials for the couple Ln° (s) $\rightleftharpoons$ Ln^{3+} (aq) + 3e^- reveal that the elemental lanthanides are powerful reducing agents, *i.e.*, oxidation of the lanthanide metals to the tripositive state takes place readily and vigorously. The gradual decrease, though very slow, in the values of E° reveals very slight decrease in chemical activity from one element to the next one.

(ii) The enhanced stabilities associated with the empty, half-filled and completely filled 4f -orbitals is also revealed by these values. Thus ce^{4+} $(4f^0)$ is much less readily reduced to tripositive ion, Ce^{3+} $(4f^1)$ than Pr^{4+} ion $(4f^1)$. Furthermore, the $4f^7$ species (*e.g.*, Eu^{2+} ion) and the $4f^{14}$ species (*e.g.*, Yb^{2+} ion) are the weakest reducing agents of the di-positive species.

(iii) The values of E° for couple Ln^o (s) $\rightarrow$ Ln^{3+} (aq) $+3e^-$ get decreased (with the increase of atomic number, as is evident from Table 1.3. The high values of E° have been found to be in accordance with the electro-positive character of lanthanides. The decrease in the values of E° with the increase in atomic number (*i.e.*, a decrease in the electro-positive character) is evidently inconsistent with the corresponding decrease in the ionic radius, *i.e.*, lanthanide contraction (to be described subsequently).

Chemistry of +3 State

In general, the lanthanides behave as active metals. Their electrode potential values have been found to be comparable to those of alkaline earths. All the metals act as strong reducing agents. They all get attacked by acids with the liberation of hydrogen. The metals dissolve slowly in cold but more rapidly in warm water, and evolve hydrogen. They absorb hydrogen to form interstitial hydrides.

Nearly all known anions have been reported to form the compounds with Ln^{3+} cation. These compounds are stable in solid as well as in solution state. Compounds of Ln^{3+} cation with the anions such as OH–, CO_3^{2-}, SO_4^2, $C_2O_4^{2-}$, NO_3^- etc. get decompose on heating, yield first basic salts and finally oxides. Hydrated salts that are having thermally stable anions such as F^-, Cl^-, Br^-, PO_4^{3-} etc. are generally insoluble.

Oxides, Ln_2O_3 : The oxides Ln_2O_3, get formed by heating the metal in oxygen or by the thermal decomposition of the Ln $(OH)_3$ or oxy salts like $Ln_2(CO_3)_3$ and $Ln(NO_3)_3$. The oxides have been found to resemble those of alkaline earth oxides. All the oxides are almost insoluble in water. They absorb CO_2 and H_2O form air to form carbonates and hydroxides respectively.

Hydroxides, $Ln(OH)_3$: The hydroxides get precipitated as gelatinous precipitates from aqueous solution by the addition of ammonia or dil. alkali to soluble salts of Ln^{3+} ion in solution. The precipitation of these hydroxides on the addition of an alkali or ammonia to their salts has been found to be as : Sc, Lu, Yb, Tm Er, Ho, Dy, Tb, Sm, Gd, Eu, Y, Nd, Pr, Ce, La.

These hydroxides are not amphoteric. These are definite compounds having hexagonal structure and are not merely hydrous oxides. They absorb CO_2 to form normal carbonates.

The oxides and hydroxides are basic. Their basicity gets decreased with increasing atomic number. Thus, La_2O_3 and $La(OH)_3$ are the most basic whereas Lu_2O_3 is a strong base and slakes like CaO on addition of water. The hydroxides get decomposed on heating to form oxides.

Oxy Salts : Lanthanide salts of most oxy acids (called oxy salts) like sulphates, nitrates, perchlorates, bromates etc. are also known. These are generally soluble and crystallize as hydrates. It is possible to prepare soluble oxy salts such as sulphate, nitrates and perchlorates by dissolving the oxides, hydroxides or carbonates in the appropriate oxyacids.

From the solution, hydrated salts can be crystalized out. The nitrates are often deliquesecent and crystalize with $6H_2O$. Solutions of oxy salts yield hydrated cations, $[Ln\ (H_2O)_x]^{3+}$ which tend to undergo slight hydrolysis in aqueous solution;

$$[Ln\ (H_2O)_x]^{3+}\ H_2O \rightarrow [Ln\ (H_2O)_{x-1}\ (OH)^{2+} + H_3O^+$$

The smaller is the ionic radius of Ln^{3+} ion, the greater would be the tendency of the ion, $[Ln\ (H_2O)_x]^{3+}$, to hydrolyze. Hence the tendency of the $[Ln\ (H_2O)_x]^{3+}$ ion to hydrolyze gets increased with increasing atomic number, because on passing from La^{3+} to Lu^{3+} there is a contraction in their ionic radii.

Double Salts : Lanthanide salts are known to form a large number of double salts. Some important double salts are

(i) Double nitrates like $(NO_3)_3.3M^{II}\ (NO_3)_2.\ 24H_2O$ (M^{II} =Mg, Zn, Ni, Mn) and Ln $(No_3)_3.\ 2NH_4NO_3.4H_2O$.

(ii) Double sulphates like $Ln_2(SO_4)_3.3M_2^ISO_4.12H_2O$ (M = alkali metal).

The double sulphates, $Ln_2(SO_4)_3.3Na_2SO_4.12H_2O$ where Ln =La Eu have been found to be only sparingly soluble in Na_2SO_4, while those where Ln = Gd–Lu have been found to be appreciable soluble. Thus a separation of lanthanides into two groups: *Cerium group* having La_{57}–Eu_{63} lanthanides and *Yttrium group* having Gd_{64}–Lu_{71} lanthanides becomes possible. As the double salts crystalize well, these are used for separating the rare earths from one another.

Halides, LnX_3 : Fluorides get precipitated by the addition of HF or a soluble fluoride to a Ln^{III} salt solution. The fluorides particularly of heavier lanthanides, are sparingly soluble in HF because of the formation of fluoro complexes.

It is possible to prepare the anhydrous chlorides by the direct combination of the elements on heating. These are best prepared by heating the oxides with carbonyl chloride ($COCl_2$) or NH_4Cl.

$$Ln_2O_3 + 3COCl_2 \rightarrow 2LnCl_3 + 3CO_2$$

$$Ln_2O_3 + 6NH_4Cl \xrightarrow{300\times} 2LnCl_3 + 3H_2O + 6NH_3$$

It is not possible to prepare the anhydrous chlorides from the hydrated chlorides, because these lose *HCl* on heating to give the oxychlorides, (LnOCl), more readily than they lose H_2O ($ScCl_3.\ 3H_2O$ and $CeCl_4.2H_2O$, however, give Sc_2O_3 and CeO_2). The chlorides are non-volatile, deliquescent solids soluble in H_2O and alcohol. The hydrated chlorides are obtained by dissolving the oxides or carbonates in HCl and concentrating the solution to crystallizing, point. They crystalize from solution usually as hexahydrates, $LnCl_3.6H_2O$.

Bromides and iodides are more or less similar to the chlorides. Iodides of the first few lanthanides are orthorhombic while those of the remaining lanthanides are hexagonal.

Carbonates, $Ln_2(CO_3)_3$: It is possible to prepare the normal carbonates by passing CO_2 into an aqueous solution of $Ln(OH)_3$. They are also prepared by adding Na_2CO_3 solution to Ln^{III} salt solution. The carbonates are insoluble is H_2O, but dissolve in acids with liberation of CO_2 and forming Ln^{III} salts.

Phosphates and Oxalates : These are also insoluble in water. All lanthanides get quantitatively precipitated as oxalates from Ln^{3+} solution containing C_2O4^{2-} ion. The precipitate on drying and ignition yields Ln_2O_3.

ION EXCHANGER ON MACRO SCALE SEPARATION

It was observed by *Powell, spedding* and their co-workers that 5% ammonium citrate at pH 2.5-3.2 which proved useful on micro (tracer) scale spread each lanthanide through a macro column in such a manner that extensive overlapping of the characteristic bell-shaped electrical elution curves (Fig. 1.5) could be prevented only by excessive increase in column length. They tried 0.1% ammonium citrate at pH 5.0–8.0 where only the chelate $[Ln\ (cit)_2]^{3-}$ is significant and succeeded in the swift separation by developing sharply defined rectangular bands of minimum overlap, Fig. 1.6 Both constraining reactions were swift and complete and the proton proved to be a suitable adsorbed (retaining) ion. Thus kilogram quantities of better than 99.92% pure lanthanides were separated, However, due to its low total capacity these conditions are still commercially unattractive.

Aminopolycarboxylates give 1 : 1 complexes of greater stability over wide range of pH, as well as larger differences in stability between adjacent lanthanides. Ethylenediamine tetracetate (EDTA) has also been employed as this complexing agent shows an average separation factor of 2.38 for adjacent lanthanides.

Solvent Extraction

The foundation for the success in these methods was laid by the investigations of *W. Fischer* and his co-worker (1937) who showed the small differences in the distribution of the trichlorides between water and partially miscible alcohols, ketones or ethers. However it was in 1949 that *Warf* observed that Ce^{4+}can be effectively separated separated from the tripositive lanthanide ions by extraction from aqueous nitric acid solution by tri-*n*-butyl phosphate. Now the method is being used on the micro as well as macro scale.

(i) Tri-n-butyl phosphate extraction : It has been observed that in the nitrate medium where extractions with tri-*n*-butyl phosphate (TBP) are most successful, an equilibrium of the following type where (org) stands for the non-aqueous phase is attained.

$$Ln^{3+}{}_{(aq)} + 3NO^{-}_{3'(aq)} + 3TBP_{(org)} \rightleftarrows Ln(NO_3)_3(TBP)_{3(org)} \quad ...(14)$$

The distribution between the phases is expressed as :

$$\lambda = \frac{C_{Ln(No3)3(TBP)3(org)}}{C_{Ln^{3+}(aq)}} \quad ...(15)$$

and for two lanthanides, the separation factor is

$$\alpha = \frac{\lambda^1}{\lambda}$$

$$= \frac{\left[C_{Ln^1(No_3)_3(TBP)_3(org)} \times C^{3+}_{Ln}(aq)\right]}{\left[C_{Ln(NO_3)_3(TBP)_3(org)} . C^{3+}_{Ln}(aq)\right]} \qquad ...(16)$$

Peppard found an average separation factor of $\alpha = 1.5$ for adjacent lanthanides in 15.8 M nitric acid 100% pure T.B.P. systems.

This type of distribution equilibrium for the reaction in (14) is expressed as

$$K = \frac{\left[C_{Ln(No_3)_3(TBP)_3(org)}\right]}{\left[C^{3+}_{Ln(aq)}][c^{3-}_{NO_3}(aq) C^3_{TBP}(Org)\right]} \qquad ...(17)$$

By adding equations (15) and (16) and simplifying we obtain,

$$1 = KC^3_{NO^-_{3(aq)}}]\,[C^3_{TBP(Org)}] \qquad ....(18)$$

The authenticity of equation (18) is confirmed by Fig. 1.7 in which $\log_{10} \lambda$ is plotted against molar concentration of NO^-_3 at constant concentration of TBP and a straight line with slope is obtained above 10M nitric acid concentrations. The graph also shows that extent of extraction increases with decreasing ion radius.

(ii) Paper Chromatography : Cerrai and Testa were able to separate some rare-earths by ascending paper chromatography, using CRL/I papers treated with 0.1M dil. (2-ethylhexyl) orthophosphoric acid (HDEHP).

Under the same conditions an increase in R_F value is noted with decreasing atomic number, Z, of the rare earths. For a given element the R_F value gets increased with the molarity of eluting acid (HCl) and with temperature, but decreases with the concentration of HDEHP. The R_F values for 1M HCl eluant are as follows :

Ln^{3+} →	La	Ce	Pr	Sm	Eu	Gd	Dy	Er	Yb	Y	Sc
R_F →	0.92	0.90	0.76	0.64	0.51	0.12	0.06	0.03	0.03	0.07	0.0

Latter, Cerrai and Triulzi employed a mixture of 3M HNO_3 and 80% CH_3OH as solvent on Whatman cellulose paper DE–20 for separating La, Ce, Pr, Nd, Sm and Gd. Whatman DE–20 is a cellulose anion exchanger, having amine groups bonded to the cellulose matrix.

It was found that a mixture of HCl and various aliphatic alcohols can successfully be used as eluant for the separation of rare-earths by paper chromatography (Whatman No. 1).

(iv) Thin layer chromatography : By using this technique **Daneels** has separated Sm, Eu, Gd and Tb using Silica gel H. Radioactive tracers were used to detect the rare-earths. The R_F values for Sm, Eu, Gd and Tb have been found to be 0.54, 0.29, 0.73 and 1.00 respectively.

PREPARATION OF THE METALS

Due to the ease of oxidation of these metal it is not a simple task to obtain the elements in the metallic state.

The following methods may be used for the preparation of metals:

(i) **Thermal decomposition of lanthanide-mercury amalgam :** Using a mercury cathode an amalgam of the lanthanides may be obtained by the electrolysis of aqueous solutions. The amalgam thus formed is thermally decomposed to obtain the metal in a finely divided state, but complete elimination of mercury is difficult.

(ii) **Electrolysis of fused chlorides :** The anhydrous lanthanide (+3) chlorides are first of all fused with sodium, potassium of calcium chloride by external heating then the molten mass is made to electrolyse in an iron pot or in a refractory lined pot which serves as the anode.

Graphite cathodes are used. The molten metal collects as a pool at the bottom of the pot from where it is withdrawn and cast into ingots.

This method does not yield the metals in high degree of purity required in modern technology.

(iii) **Metallothermic reduction of the salts :** In this method the anhydrous fluorides and chlorides of lanthanides are made to reduce with calcium in a tantalum container at about 1000° C. Calcium is volatilised from the lanthanide-calcium alloy. This method has been found to be successful for all the lanthanides except Sm, Eu and Yb, because these metals form stable divalent fluorides.

However, these metals can be obtained by reducing their oxides with lanthanum metal at 1400°C in vacuum and distilling out Sm, Eu and Yb. These metals are more volatile than lanthanum.

The metals can be purified by vacuum distillation (pressure ~ 10–5 mm Hg) in tantalum apparatus.

APPLICATIONS OF LANTHANIDES

For the sake of convenience only, the applications of lanthanides may be classified and studied as under :

1. In Atomic Energy (Nuclear Applications)

(a) Some of the lanthanides are able to stop or absorb neutrons and are used in atomic reactors to control the rate of fission. Long rods made of these materials are introduced into the core of a reactor before it is fueled. After the addition of fuel, if some of the rods are withdrawn the fission begins, and with the removal of more rods the fission rate accelerates. If all the control rods are introduced in the core, the fission stops. Gadolinium has the largest known nuclear cross-section, or neutron stopping capability, samarium comes next and is followed by europium and dysprosium.

(b) A number of lanthanide isotopes have desirable properties for special applications. They have some potential uses in atomic batteries as gamma ray or X-ray sources; as radioactive materials for treatment of cancer and in tracer studies.

(c) Lanthanides also find some application as diluents of atomic of atomic fuels, materials to contain fuels and materials to separate undesirable fission products from atomic fuels.

2. Commercial Uses

(a) As metals and alloys : They have hardly any use in the elemental state. They are used mostly as alloys (Misch metals) containing predominantly Ce (30-50%) along with small quantities of other cerium group metals and non-lanthanides. The typical composition of Misch metal is cerium 45-50%; La 22-25%; Nd 18%, Pr 5% Sm 1% and smaller quantities of other lanthanides. Misch metal has strong reducing property. It is a very good scavenger for oxygen and sulphur in several metallurgical operation.

Magnesium alloys with Misch metal (3%) and Zr (1%) possess high strength nd resistance and are useful in jet engines. Misch metals also increase the resistance of nickel alloys and working ability of stainless steel and vanadium. When alloyed with 30% iron, it is pyrophoric and hence useful for lighter flints.

(b) In the form of compounds : The CeO_2 is used for grounding and polishing optical glass, L_2O_3 is added to camera lenses to reduce the chromatic aberration (the spreading of colours as they pass through the lens). As fluorides about one-fourth of lanthanides is used in cored carbons for improving the intensity and colour balance, *i.e.,* uniformity in arc search lights and motion picture projectors. Lanthanide oxides are dissolved in glass to impart beautiful colours to glass windows and glass vases.

3. As Catalysts

The oxides of lanthanides are used in hydrogenation, dehydrogenation and oxidation of various organic compounds, *e.g.,* their anhydrous chlorides in polyesterification processes, and the chlorides and serum phosphate in petroleum cracking. They have lot of scope in catalysis, since heterogenous catalysis is usually characterized by unpaired electrons and variable oxidation states etc.

4. Magnetic and Electronic Applications

Their paramagnetic and ferro magnetic properties find application in this field also. They find use in micro wave devices due to the low electrical and eddy current losses of the ferro magnetic garnets, $3Ln_2O_3 . 5Fe_2O_3$. Some compounds of these elements have potential use as semi-conductors or thermoelectrics. Due to large dielectric constant and small temperature coefficients of capacitance of their titanates and stannates, they are useful ceramic capacitators.

Comparison of Lanthanides and Actinides

There are many points of similarity between actinides and lanthanides. In both these series $(n–2)$ *f*-shells are progressively filled. There is actinide contraction similar to lanthanide contraction. Like lanthanide ions, many actinide ions coloured and show paramagnetic behaviour. They also differ in some respects. Actinides have a far greater tendency to form complex than lanthanides. The difference is due to relatively lower binding energies of actinides. $5f$ level becomes progressively lower in energy as compared to 6d level as the atomic number increases. So in this series the energies of $5f$, $6d$ and $7p$ are comparable over a range of atomic number (U to am) and these orbitals also overlap each other in space.

Various points of similarity and difference are summarized below.

Points of Similarity

(i) Both show +3 oxidation state.

(ii) f-orbitals of (n-2) shell are progressively filled up in both or them.

(iii) A contraction in atomic size is observe in both cases with increased in atomic number.

(iv) Due to f-f transition, the absorption spectra of the elements of both the series give sharp line-like band spectra.

(v) The electronegativity values of elements of both the series are low.

(vi) These elements are quite reactive.

(vii) The nitrates, perchlorates and sulphates of trivalent elements of both series are soluble.

(viii) The carbonates and hydroxides of trivalents of both the series are insoluble.

(ix) Members of both the series show ion-exchange behaviour.

Comparison of 3d-Elements (Transition series) with 4f-elements (inner transition series)

In both these cases ($3d$ and $4f$ series) the elements have incompletely filled inner orbitals. These elements show many properties in common but they also show some difference in their properties. The reason is that in *d*-block elements the electrons are entering penultimate (last but one) shell whereas in *f*-block elements the electrons are entering anti-penultimate (last but two) shell. Therefore, d-orbitals project out and electrons residing in them get strongly influenced by the nature of the ligands which are surrounding the metal atom or ion, *i.e.,* d- electrons can interact strongly with their surrounding. On the other hand, *f*-orbitals in lanthanides get deeply buried in the atoms or ions and fail to come in contact with the surrounding ligands. Therefore, the electrons residing in these orbitals get effectively screened from the surroundings by the overlapping orbitals nd therefore the chemistry of lanthanides (and actinides) would be very similar to each other. Thus, the differences in their properties arise due to :

(i) Different number of d-and f-orbitals.

(ii) Difference in energies of (n–1) d-orbitals and (n–2) f-orbitals. (n–1) d > (n–2) f.

(iii) Proximity of (n–2) f-orbitals to the nucleus as compared to that of (n–1) d-orbitals.

(iv) (n–2) f-orbitals can participate in bonding only with great difficulty as compared to (n–1) d-orbitals.

Important Points of differences are :

(i) **Size :** The atomic size of the elements of d-block series are smaller as compared to the atomic size of the f-block elements.

(ii) **Ionization energy :** The values of first ionization energy of d-block elements are higher as compared to those of f-block elements.

(iii) **Oxidation states :** Much greater variability is observed in oxidation state of d-block elements as compared to f-block elements particularly lanthanides.

(iv) **Absorption spectra :** The absorption spectra of f-block elements are narrow bands whereas d-block elements form elements particularly lanthanides.

(v) **Basic nature :** The oxides and hydroxides of f-block element are more basic than those of f-block elements.

(vi) **Oxidation potential :** The oxidation potential of f-block elements are comparatively much higher as compared to those of d-block elements.

(vii) **Complex formation tendency :** This tendency is much more in case of f-block elements in comparison to d-block elements.

(viii) **Coordination number :** The coordination numbers shown by lanthanides are higher than those shown by transition elements.

(ix) **Magnetic moments :** The maximum magnetic moment in f-block elements correspond to 7 unpaired electrons and in case of d-block elements to 5 unpaired electrons. It is easier to explain the magnetic properties in case of d-block elements but very difficult to explain the magnetic properties of f-block elements.

The experimental magnetic moments are usually lower than the values calculated by using Russel-Saunders coupling. This appears to be due to ligand field effect similar to those operating in the d-transition series. It is now quite certain that 5f-orbitals can participate to some extent in covalent bonding so that ligand effects are to be expected. For the ions Pu^{3+} and Am^{3+}, the same phenomenon is noted as for Sm^{3+} and Eu^{2+}.

LANTHANIDES (4F-BLOCK SERIES)

The elements in which the additional electron enters $(n-2)$ *f*-orbitals are termed as inner transition elements. These are so called because literally speaking they constitute transition series within transition series (*d*–block elements). The valence-shell electronic configuration of these elements can be represented as $(n-2) f^{0},{}^{2}, ...{}^{14} (n-1) d^{0},{}^{1}, {}^{2} ns^{2}$. These are also termed as *f-block elements* because the extra electron enters f-orbitals which belong to $(n-2)$th shell. In addition to incomplete *d*-subshell, their *f*-subshell is also incomplete.

As there are seven *f*-orbitals in a given shell, it means that fourteen electrons can be occupied in a given *f*-block *i.e.,* a particular *f*-block series should have 14 elements.

It is possible to sub-divide *f*-block elements into two series depending upon the nature of the *f*-orbital of the antipenultimate shell (4*f* or 5*f*) in which the differentiating electron enters.

(a) 4*f*-Block Series, Lanthanides or Lanthanones

In these elements the differentiating electron enters 4*f*-orbitals. This series is having lanthanum (Z = 57) and the next fourteen elements (Z = 58 to 71). If this definition is strictly followed, only thirteen elements from Ce_{58} $(4f^{1}5d^{1}6s^{2})$ to Yb_{70} $(4f^{13}\,5d^{1}\,6s^{2})$ should be the members of this series because lanthanum La_{57} $(5f^{0}5d^{1}\,6s^{2})$ and lutetium Lu_{71} $(4f^{14}\,5d^{1}\,6s^{2})$ are having either completely empty or completely filled *f*-orbital. However, the fourteen elements from Ce_{58} to Lu_{71} are generally regarded as 4*f*-block elements. As the number of electrons in the outermost, as well as in the penultimate shells remains the same, the fourteen elements resemble one another.

The 4*f*-block elements are also called lanthanides, lathanones or rare-earths. The first two names are given because of their strong resemblance to lanthanum. The 4*f*-electrons in these elements are well shielded and are not available for chemical bonding, whereas in transition elements the added *d*-electrons are available for chemical bonding. Because of the this *lanthanides* show a different behaviour than that of transition elements (*d*-type). Moreover, the properties of Sc (At no. 21), and lanthanum (At. no. 57) are different from those of lanthanides because of the absence of 4*f*-electron in them.

Klemm has divided lanthanides into two groups of seven elements each. These are

(i) form Ce (58) to Gd (64), and

(ii) from Tb (65) to Lu (71).

In case of (i) the half-filling of 4f-orbitals takes place and in case of (ii) the pairing of electrons takes place.

The name rare earth was given to them because they were originally extracted from oxides for which ancient name was earth and which were considered to be rare. The term rare-earth is avoided now because many of these elements were divided into three groups.

(i) **Cerium group :** It includes Ce, Pr, Nd and Sm in addition to La; Pm is not included in this family. The double sulphates of these elements with K_2SO_4 are soluble in water but these are insoluble in cold saturated solution of H_2SO_4.

(ii) **Terbium family :** Eu, Gd and Tb are included in this family. The double sulphates of these elements with K_2SO_4.

(iii) **Yttrium family :** Dy, Ho, Er, Tm, Y, Yb, Lu are included in this family. The double sulphates of these elements with k_2SO_4 are soluble in cold saturated solution of K_2SO_4.

The name lanthanides has been derived from lanthanum which is the prototype of lanthanides.

(b) 5*f*- Block Elements, Actinides or Actinones

In these elements the differentiating electron enters 5*f*-orbitals. This series includes fourteen elements from thorium (Z = 90) to lawrencium

(Z = 103) and described after lanthanides. The name actinide has been derived from actinium, the very first member of the series.

It is interesting to observe that the *f*-block elements (lanthanides and actinides) are kept outside the body of the periodic table. The reason for this is attributed to the remarkable similarities among the chemical properties of the lanthanides and also among the various members of actinides.

The similarities in properties, in turn, are ascribed to the similar electronic configurations of the outermost shell. These elements differ only in the number of *f*-electrons which are not involved in chemical bonding (difference from *d*-block elements in which differentiating *d*-electrons are involved in chemical interaction).

In case, if we place these elements inside the body of the periodic table, the very purpose of the periodic law (*i.e.,* the elements arranged in the periodic table must show gradation in their physical and chemical properties) would be marred.

General characteristics of lanthanides

Although, La_{57} and the following elements up to Lu_{71} have been found to resemble each other in their properties, lanthanum itself has been studied in d-block elements and not f-block because here f-orbital is having no electron.

The *f*-orbital actually starts filling at cerium, Ce (Z = 58) and is completely filled at lutetium, Lu (Z = 71) and hence elements from cerium (Z = 58) to Lu (Z = 71) have been actually grouped as 4f-block elements. Further as these fourteen elements follow lanthanum, these are also termed as *lanthanides or lanthanones.*

ELECTRONIC CONFIGURATION OF LANTHANIDES

It is not possible to know the electronic configurations of lanthanides with certainty. The probable reason for this is that the 5d and 4f electrons are so close in energy that it is not always possible to ascertain whether the electron is entering 5d-orbital or 4f-orbital. The above electronic configuration can be understood by considering the electronic configuration of La (at. no. 57), the element that proceeds this series and has the following electronic configuration.

$$La = 57\ 1s^2 2s^2 2p^6 3s^2 3p^6 d^{10} 4s^2 4p^6 4d^{10} 5s^2 5p^6 5d^1 6s^2$$

or [Xe]$5d^16s^2$

As we proceed further to at. no. 58 the energy 4*f*- level falls below that 5*d*-level and so the electrons now enter inner 4*f*-orbitals. The seven *f*-orbitals can accommodate 14 elections thus an inner transition series of 14-elements results. These are the elements from Ce to *Lu*. Because these elements follow lanthanum (57) they are known as lanthanides and represented as Ln collectively. The second view is that the electronic structure of lanthanides involved the movement of single 5*d* electrons into 4*f*-sub shell and thus it does not apply only when 4*f*-orbitals are completely half filled as in case of *Gd* (at. No. 64) or when 4*f*-orbitals are completely filled as in case of *Lu* (at. no. 71).

It should be clearly noted that in all these elements the electron enters the 4*f*-orbitals of the (*n*–2) shell. These orbitals lie below 5*s*, 5*p* and 6*s* electrons which are present in these atoms. Because of similarity in the electrons present in outer shell, diversity being only in 4*f*-electrons these elements show a very close resemblance in the chemical properties. The small differences observed in the properties of these inner transition elements can be accounted for by the decrease in atomic radii that is observed as we move from *La* to *Lu*.

ELEMENT OF GROUP IIIB

This group compirses of the elements, namely scandium, yttrium, lanthanum, and actinium. These are d-block elements having the electronics configuration, $(n-1)d^1\ ns^2$, Table 3.

Table 3

Element	*Symbol*	*At. No.*	*Electronic configuration*	*Oxidation state*
Scandium	SC	21	[Ar] $3d^14s^2$	3
Yattrium	Y	39	[Kr] $4d^15s^2$	3
Lanthanum	La	57	[Xe] $5d^16s^2$	3
Actinium	Ac	89	[Rn] $6d^17s^2$	3

These four elements are sometimes grouped with the fourteen lanthanons and called collectively the 'rare earths'. This misnomer because the scandium group include d-block elements and the lanthanons are f-block elements. Also, the scandium group is by no means rare except for actinium which is radioactive.

General characteristics : Some of the important characteristics of the scandium group are given in Table 46.2.

1. **Physical state :** There are metals having high melting points. However, the melting points decrease on descending the group from Sc to Ac.
2. **Size :** The covalent radii of the elements increase regularly on descending the group as in the s-block. Similarly, the ionic radii increase from Sc^{3+} to Ac^{3+}.
3. **Ionization energies :** They decease regularly on descending the group indicating that the electropositive character increases on moving down the group.

   Further, the sum of the first three ionisation energies for scandium is a little less than the sum for aluminium, and the properties of scandium are similar in some ways to those of aluminium.

Table 4

Element	*Covalent*	*Ionic*	*Ionisation*			*M.P.(K)*	*Pauling's*	*Abundance*
		radius,	*radius*			*energies*		*Electro*
earth's crust,						*(Å)*	*(M^{3+}, Å)*	*(kJmol^{-1})*
negativity		*ppm*						
			1st	*2nd*	*3rd*			
Sc	1.44	0.70	631	1235	2393	1812	1.3	5
Y	1.62	0.90	616	1187	1968	1782	1.2	28
La	1.69	1.06	541	1100	1852	1193	1.1	18
Ac	1.11				~1050		1.1	trace

4. **Oxidation state :** These elements show an oxidation state of + 3 and occur as M^{3+} ions in solution.
5. **Colourless ions and salts :** The trivalent ions are formed by the removal of the two s and one d electron, thus, giving a d^{o} configuration and consequently, the ions and their compounds are colourless diamagnetic and colourless.
6. **Reactivity :** They are quite reactive metals and their reactivity increases with increased size from Sc to Ac. This trend is evident from the following chemical properties :

(a) *Action of air :* All these metals tarnish in air. However, Y forms a protective oxide coating in air, which makes it unreactive.

(b) *Action of oxygen :* They burn in oxygen forming oxides, M_2O_3.

$$4Sc + 3O_2 \rightarrow 2Sc_2O_3$$

(c) *Action of water :* They are highly electropositive metals and liberate hydrogen slowly with cold water and more rapidly on heating forming basic oxides and hydroxides.

$$2La + 6H_2O \rightarrow 2La(OH)_3 + 3H_2$$

$$\downarrow -H_2O$$

$$LaO.OH$$

Basic oxide

(d) *Action of halogen :* The metals react with halogen to form trihalides.

(e) *Action of hydrogen :* When heated to 300°C, they react with hydrogen to form compounds of formula MH_3. These compounds are not quite stoichiometric. Also, their exact composition depends on the temperature and the pressure of the hydrogen.

(f) *Action of carbon :* They do not react with carbon. However, only scandium forms a carbide ScC_2 by heating the oxide with carbon at 1273 K.

$$Sc_2O_3 + 7C \rightarrow 2ScC_2 + 3CO$$

Earlier, this carbide was considered to be divalent Sc^{2+} (C = C)2^- Recently, it has been shown by magnetic and conductance measurements that it contains Sc^{3-} ions, C^{2-}_2 ions and free electrons. The carbide reacts with water, liberating acetylene.

$$ScC_2 + H_2O \rightarrow C_2H_2 + ScO.OH$$

(g) *Tendency to form complexes :* Tendency to form complexes decreases with increase in size from Sc to Ac. Scandium is the smallest ion and therefore forms complexes more readily than the elements. Examples of such complex ions are

$[Sc(OH)_6]^{3-}$ and $[ScF_6]^{3-}$ (formed by dissolving ScF_2 in HF) which are octahedral in shape.

$$ScF_2 + 3NH_4F \rightarrow [ScF_6]^{3-} + 3NH_4^+$$

7. Nature of Compounds

(a) **Nature of oxides and hydroxides :** Basic properties of the oxides and hydroxides increase on descending the group. $Sc(OH)_3$ appears not to exist as a definite compound, but $ScC(OH)_2$ is well established and is amphoteric like $Al(OH)_3$. $Y(OH)_3$ is basic in nature while $La(OH)_3$ is a strong enough base to liberate NH_3 from ammonium salts. Because of their basicity, the oxides and hydroxides form salt with acids. $Y(OH)_3$ and $La(OH)_3$ react with carbon dioxide.

$$Sc(OH)_3 + 3HCl \rightarrow ScCl_3 + 3H_2O$$

$$2Y(OH)_3 + 3CO_2 \rightarrow Y_2(CO_3)_3 + 3H_2O$$

As the oxides (and hydroxides) are either amphoteric or weak bases, their salts on heating decompose to oxides.

$$2Y(OH)_3 \xrightarrow{\text{Heat}} Y_2O_3 + 3H_2O$$

$$Y_2(CO_3)_3 \xrightarrow{\text{Heat}} Y_2O_3 + 3CO_2$$

$$Y_2(SO_4)_3 \xrightarrow{\text{Heat}} Y_2O_3 + 3SO_3$$

$$Y_2(NO_3)_3 \xrightarrow{\text{Heat}} Y_2O_3 + 3NO_2$$

(b) **Nature of halides :** Fluorides are insoluble and others are deliquescent and very soluble like $CaCl_2$. Chlorides crystallise as hydrated slats. When these are decomposed by heating, the hydrated scandium chloride gives the oxide rather than the anhydrous chloride and the others giving oxyhalides.

$$ScCl_{3.}H_2O \xrightarrow{\text{Heat}} Sc_2O_3 + HCl$$

$$YCl_{3.}H_2O \xrightarrow{\text{Heat}} YOCl + HCl$$

$$LaCl_{3.}H_2O \xrightarrow{\text{Heat}} LaOCl + HCl$$

(c) **Natur of hydrides :** These are slat-like (ionic) hydrides containing the H– hydride ion. These react with water, liberating hydrogen.

SYMBOL OF THE SCANDIUM

Scandium occurs in very low concentration in silicate rocks in the earth's crust and is about as abundant as beryllium and the rate earths about (5p.p.m.), although one mineral, *thortveitite* contains 30 to 40% scandium oxide.

Extraction

It is not an easy task to extract scandium from its compounds. The oxide is highly stable, even more than that of alumina. Therefore, thermit process cannot be employed for its extraction from its oxide. Further, the electrolysis and reacts with water to liberate hydrogen.

Metallic scandium may be prepared by first treating the oxide with anhydrous HF to form ScF_3. The salt is reduced with calcium metal in a tantalum crucible at 1873K in a purified argon atmosphere. The scandium is obtained as a solid in-got which has a slight yellowish colour and contains 3% or more tantalum present as pure tantalum dendrites. Careful sublimation in a high vacuum at 1773 K produces good quality metal.

Alternatively, scandium is extracted by electrolysing fused scandium chloride containing a small amount of sodium chloride which lowers its melting point, Zinc is used as the cathode. On passing electric current, scandium is deposited on the zinc cathode in the form of Zn– Sc alloy. Zinc is volatilised off at low pressures when scandium is left behind in reasonably pure state.

Properties

It is a silvery white metal having a metallic lustre. In air, it is tarnished due to the formation of its oxide, Sc_2O_3 on the surface. Its melting point is 1812 K and boiling point –3000K.

In the metallic state, scandium has three conduction electrons ($4s^2 3d^1$) and loses these when dissolved to give colourless aqueous solutions, with scandium in a oxidation state of + 3. Compounds with higher oxidation states are not known. Also, the + 2 state is not stable.

Comparison with Aluminium

The sum of the first three ionization energies for scandium is a little less than the sum for aluminium, and the properties of scandium are similar in some ways to those of aluminium. This is shown as below:

1. Formation of M^{3+} ions : The outer electronic configuration of scandium is $3d^14s^2$ while that of aluminium is $3s^2p^1$. Both give trivalent ions, Sc^{3+} and Al^{3+} in solution by losing their three outer electrons. *i.e.,* two s and one d electron in Sc while two s and one p electron in Al. Both these ions have a noble gas configuration :

$$Sc : [Ar]\ 3d^14s^2 \ ;\ Sc^{3+} : [Ar]$$

$$Al : [Ne]\ 3s^24p^1 \ ;\ Al^{3+} : [Ne]$$

2. Colourless and diamagnetic compounds : Both Sc and Al form colourless and diamagnetic compounds.

3. Reaction with water : Both react with water, liberating hydrogen.

$$2Sc + 3H_2O \rightarrow Sc_2O_3 + 3H_2$$

$$2Al + 3H_2O \rightarrow Al_2O_3 + 3H_2$$

4. Nature of oxide : Both form amphoteric oxides. Both the oxides are in soluble in water.

5. Formation of double sulphates : Both form double sulphates:

$$K_2SO_4.Sc_2(SO_4)_3.xH_2O ; K_2SO_4.Al_2(SO_4)_3.24H_2O$$

Scandium differs from aluminium in the following ways :

(i) The ionic radius of Sc^{3+} is much larger than that of Al^{3+}.

(ii) $ScCl_3$ differs from $AlCl_3$ since $ScCl_3$ is monomeric while $AlCl_3$ is dimeric *i.e.,* $(AlCl_3)_2$. Also $ScCl_3$ does not show Friedel Crafts catalytic properties while $AlCl_3$ shows this property.

Comparison with Lanthanides : Scandium has been found to resemble lanthanides in the formation of insoluble salts like oxalate, phosphate, carbonate and fluoride.

As Sc^{3+} ion (0.81 Å) has a smaller radius than that of lanthanides (in the range of 0.85-1.06Å), it differs appreciably from lanthanides in the following ways :

1. **Nature of the oxide :** The scandium oxide has been found to be less basic than the oxides of lanthanides.
2. **Nature of chloride, nitrate and sulphate :** The $ScCl_3$ is volatile and sublimes at a much lower temperature than lanthanide chlorides. Scandium nitrate, $Sc(NO_3)_3.4H_2O$ is more easily decomposed than lanthanide nitrates. Scandium sulphate,

$Sc_2(SO_4)_3.5H_2O$ is soluble in cold as well as hot water while the lanthanide sulphates are not.

3. **Double salt :** Scandium is known to form a double salt of composition, $K_2SO_4.Sc_2(SO_4)_3.nH_2O$. However, lanthanides do not form such salts.
4. **Formation of thiocyanate :** When treated with KCNS. Scandium salts gives a yellow ppt. of $Sc(CNS)_3$. However, salts of lanthanides do not give such reaction.
5. **Hydrolysis :** Due to its smaller size, scandium ion has a greater tendency to hydrolyse than the lanthanide ions.
6. **Formation of complex :** Due to its smaller size, scandium ion forms complexes more readily than the lanthanide ions.
7. **Polarising power :** Due to its smaller size, scandium ions have greater polarising power and greater tendency to introduce some electrovalent character in its compounds than the lanthanide ions.

Compounds of Scandium

Scandium is known to form compounds like oxide, halide and oxyhalides in which its oxidation state is + 3.

1. Scandium Oxide, Sc_2O_3 : It is obtained readily by heating the salts such as scandium carbonate and scandium nitrate.

$$Sc_2(CO_3)_3 \xrightarrow{\text{Heat}} Sc_2O_3 + 3CO_2$$

$$4Sc(NO_3)_3 \rightarrow Sc_2O_3 + 12NO_2 + 3O_2$$

Scandium oxide is a white powder. It is more basic than aluminium oxide but less basic than calcium oxide. It absorbs carbon dioxide from the air.

It has the Mn_2O_3 structure.

2. Scandium hydroxide : $Sc(OH)_3$ does not appear to exist as a definite compound but ScO.OH is well established and is amphoteric like $Al(OH)_3$. Scandium hydroxide is thrown down as a gelatinous precipitate on the addition of an alkali to a solution of scandium chloride.

$$ScCl_3 + 3NaOH \rightarrow ScO.OH + 3NaCl + H_2O$$

The lattice of scandium hydroxide is similar to the of Y.AlO.OH.

3. Scandium hydride : Scandium reacts with hydrogen on heating to 573K forming compound of formula ScH_3 which is not stoichiometric. Its exact composition depends upon the temperature and the pressure of the hydrogen. Scandium hydride is salt-like hydride containing the H^- ion. It reacts with liberating hydrogen.

4. Scandium carbide ScC_2 : It is obtained by heating the oxide with carbon at 1273 K.

$$Sc_2O_3 + 7C \rightarrow 2ScC_2 + 3CO$$

It is an acetylide because it reacts with water to liberate acetylene.

$$ScC_2 + H_2O \rightarrow C_2H_2 + ScO.OH$$

Earlier this carbide was considered to be divalent Sc^{2+} $(C \equiv C)^{2-}$. Recently It has been shown by conductivity and magnetic measurements that it contains Sc^{2+} and C_2^{2-} ions.

5. Complexes : As scandium is the smallest ion, it forms complexes more readily than the other elements. For example, scandium fluoride dissolves in the presence of excess HF or NH_4F give $[ScF_6]^{3-}$ which is octahedral in shape.

$$ScF_3 + 3NH_4 \rightarrow [ScF_6]^{3-} + 3NH_4^+$$

YTTRIUM

Occurrence

It is present in the earth's crust to the extent of 40 ppm. Yttrium is present as a phosphate (*xenotime*), silicate (*gadolinite*), etc.

Extraction

In processing the ores, an electromagnetic concentration is followed by a high-temperature sulphuric acid digestion and aqueous leach. The resulting solution is leached on ion-exchange resin beds which are elated with an EDTA solution that is buffered with ammonia and also contains copper ion which serves to prevent precipitation of the insoluble EDTA in the resin bed. The rare earths come off in inverse order of their atomic numbers; yttrium comes between holmium and dysprosium.

Properties

In aqueous solution, yttrium loses three conduction electrons of the metallic form ($5s^2 4d^1$) to give the colourless ion.

Uses : Yttrium is finding much usage in the colour television industry where Y_2O_3 – Er and YVO_4 – Er phosphorus provide the vivid red colour. Small amounts of yttrium improve the corrosion resistance of chromium, niobium, tantalum and molybdenum.

LANTHANUM

Electronic configuration of lanthanum (atomic number 57) has been KLM $4s^2 4p^6 4d^{10} 5s^2 5p^6 5d^1 6s^2$. Fourteen elements following it are called lanthanons or lanthanides. In these 4f orbitals are being filled by electrons. The idealised outer electronic configuration of lanthanons has been $4f^{1-14}\ 5d^1 6s^2$ where n varies from 0 to 14. However, there are some discrepancies because the half-filled and completely filled orbitals are having extra stability. Thus europeum (Eu, atomic number 63) is having the configuration $4f^7 6s^2$ instead of the expected $4f^6 5d^1 6s^2$.

OCCURRENCE AND EXTRACTION

Lanthanum is found in materials having rare earths like cerite, allanite and monazite. Cerium family members (called ceria earths) could be separated from the rest by appropriate means. Lanthanum also passes into the ceria earths. Lanthanum could be prepared after the removal of cerium by fractionation with ammonia or magnesia. Other methods used are fractional crystallisation of various double nitrates or bromates, electrolysis, extracting the roasted oxides by mineral acids and precipitation as oxalates or chromates. It could also be prepared by heating its chloride with sodium of potassium.

$$LaCl_3 + 3K \rightarrow La + 3KCl$$

Electrolysis of fused lanthanum chloride covered with ammonium chloride could be used to get the metal.

Properties

Lanthanum is a white metal like tin. However, it is harder than both lead and tin. It is sufficiently malleable and ductile to get hammered into foil and drawn into wire.

As lanthanum is an active metal, it gets tarnished rapidly even in dry air. It forms the nitride LaN when heated in nitrogen. Its carbide LaC_2 and sulphide La_2S_3 could be prepared on heating with carbon and sulphur respectively. It reacts slowly with cold water but rapidly with hot water forming the hydroxide.

$$2La + 6H_2O \rightarrow 2La(OH)_3 + 3H_2$$

In the form of filings it rapidly reacts with chlorine on heating to form the chloride, $LaCl_3$. It reacts with bromine at 490K; reaction with iodine vapour has been slow. It gets oxidised violently with nitric acid and is rapidly attacked by other acids.

Uses : It is a component of mischmetal which finds extensive use in making gas lighters and cigarette lighters. Mischmetal (an alloy containing Ce, La and some other earth elements) acts as a powerful reducing agent and finds use to reduce oxides of metals.

Compounds of Lanthanum

Lanthanum oxide or Lanthana, LaO_3 can be prepared by heating by the metal in air or igniting its hydroxide, carbonate nitrate and oxalate. It absorbs carbon dioxide from air to from $La_2(CO_3)_3$. It forms the hydroxide $La(OH)_3$ with a hissing sound when water is sprinkled over it (cf. CaO).

It exhibits some amphoteric character because on fusion with sodium carbonate or digestion with excess of caustic soda it yields *sodium lanthanate,* $Na_2La_4O_7$.

Lanthanum hydroxide, $La(OH)_3$ gets precipitated when an alkali solution or ammonium hydroxide gets added to a solution of one of its salts. It is slightly soluble in water.

It solution is alkaline towards litmus. It absorbs iodine and yields a blue colour like search. This colour gets destroyed by acids and alkalis.

Lanthanum carbonate, $La_2(CO_3)_3$ gets precipitated when an alkali carbonate has been added to a solution of lanthanum salt. It is soluble in excess of alkali carbonate solutions because of the formation of double salts like $K_2CO_3La_2(CO_3)_3.12H_2O$.

Lanthanum sulphate, $La_2(SO_4)_3$ and lanthanum nitrate, $La(NO_3)_3$ could also be prepared by the action of sulphuric and nitric acid respectively on the metal, as well as its oxide, hydroxide and carbonate. These are known to form double salts like.

$$K_2SO_4.La_2(SO_4)_3;\ 2NaNO_3.La(NO_3)_3.H_2O;$$
$$2NH_4NO_3.La(NO_3)_3.4H_2O$$

Lanthanum nitrate finds use for detection of acetate. To the acetate solution add a few drops of lanthanum nitrate followed by iodine solution and ammonium hydroxide. The mixture is heated when deep blue precipitate is obtained due to the adsorption of iodine on basic lanthanum acetate.

The halides $LaCl_3$, $LaBr_3$ and LaI_3 could also be prepared by direct action of the halogens on the metal as well as the action of corresponding hydrogen halides on the oxide, hydroxide and carbonate of the metal.

Lanthanum fluoride, LaF_3 could also be prepared as a white gelatinous precipitate by treating lanthanum acetate with hydrofluoric acid. With excess of hydrofluoric acid it yields a double salt, $2LaF_3.3HF$.

Lanthanum sulphide, La_2S_3 could also be prepared by passing the vapour of carbon disulphide over heated lanthanum carbonate, or by heating the anhydrous sulphate in a steam of hydrogen sulphide.

POSITION IN THE PERIODIC TABLE

To understand the position of these elements in the periodic table we start with the electronic configuration of Xe (at. no. 54).

Xe has the electronic configuration as follows: $1s^2 2s^2 p^6 3s^2 3p^6 3d^{10} 4s^2 4p^6\ 4d^{10} 5s^2 5p^6$

We observe that at xenon the 6*s* and 6p-orbitals become stable. Since these orbitals are more penetrating so they feel greater value of Z* and thus lowered in energy (Z* is the effective nuclear charge).

The further addition of the electrons takes place in 6*s*-orbital giving an alkali metal (Cs) and a rare earth elements (Ba). So at barium (Ba) all the three sub-shells 6*s*, 5*d* and 4*d* have very similar energies. In lanthanum (La) the electron enters 5*d*-orbital and its configuration is La_{57} $[Xe]6s^2 5d^1 4f^0$.

As we move a step further the 4*f*-sub shell becomes definitely more stable than 5*d*-sub shell and therefore the next-electron in Ce (58) enters the 4*f*-sub shell so the configuration of Ce becomes Ce_{58} $[Xe]6s^2 4f^2$.

The filling of 4f-sub shell continues till we reach Yb having the configuration Yb_{70} $[Xe]\ 6s^2 4f^{14}$

Now when 6*s* and 4f orbitals are completely filled the next electrons enters 5*d*-orbitals and so Lu (71) has the electronic configuration $[Xe]6s^2 4f^{14} 5d^1$.

Table 4 list some of atomic properties of lanthanides whereas electronic configuration of these elements are given in Table 1.1.

These elements can be given the general electronic configurations as [Xe] $6s^2 4f^{0-14} 5d^{0-1}$.

Because of similarity in electronic configuration they show similar chemical properties and are grouped at one position in the periodic table.

Oxidation States

All the elements of this series exhibit an oxidation state of +3 (Ln^{3+}). Some of these also exhibit +2 and +4 states but these states are not as stable as +3 state. Only +3 state is exhibited by *La*, *Gd* and *Lu* which is due to extractability of empty, completely half filled or completely filled 4*f*-orbitals, other oxidation states exhibited by some member of this series can also be explained on the basis of extra stability of $4f^0$, $4f^{14}$ configuration. *Eu* and Yb can acquire 2+ state as it provides them with extra stability because they now acquire f^7 and $4f^{14}$ structures respectively. However, the existence of some in +2 and +4 states can not be explained on this principle. In such cases it can be argued that the stability of these unusual states depends upon some other factors which are of significant importance. Such factors may be thermodynamic or kinetic.

The M^{4+} ions in aqueous solutions readily change into M^{+3} ion *e.g.*,

$$Ce^{4+} + e \rightarrow Ce^{3+}, E^o = 1.161 \text{ V}.$$

Thus, Ce^{4+} is a strong oxidizing agents. The Pr^{4+} and Tb^{4+} are still more powerful oxidizing agents.

Table 1.2 leads to following conclusion.

(i) Lanthanum exhibits only the +3 oxidation state which is the characteristic and the most stable oxidation state of all the lanthanides (Ln^{3+}). It is to be noted that the configuration of La^{3+} ion is that of xenon which being highly stable does not lose forming La^{4+} ion.

(ii) Gadolinium (Gd) and lutetium (Lu) are also known to form the tripositive, M^{3+} ions since the removal of three electrons yields the stable $4f^7$ (half filled) and $4f^{14}$ (completely filled) configurations respectively. Thus, any oxidation state other than +3 is not found in these elements.

(iii) In addition to the usual +3 oxidation state, few of the lathanides also exhibit +2 and +4 oxidation states. The +2 and +4 oxidation states may be ascribed to the extra stability associated with empty, half-filled or completely-filled 4f subshell. For example. Ce^{4+} has f^0; Eu^{2+} has f^7 Yb^{2+} has f^{14} and Tb^{4+} has f^7configuration.

Table 5 : Important Characteristics of Lanthanides

Element	*At. No.*	*At. Wt.*	*Densities g Cm^{-3}*	*Atomic Radius* Å*	*Ionic Radius (M^{3+}) (Å)*	*Ox. on States*	*M.P. (°C)*	*E°(volt)** $Ln^{3+}+3e^- \to Ln$*
La	57	138.91	6.174	1.877	1.045	+3	920	–2.52
Ce	58	140.12	6.77	1.825	1.030	+3,+4	795	–2.33
Pr	59	140.91	6.782	1.828	0.997	+3, +4	919	–2.47
Nd	60	144.24	7.004	1.821	1.000	+2, +3	1020	–2.45
Pm	61	147.00	–	1.810	0.980	+3	1080	–2.42
Sm	62	150.40	7.536	1.802	0.958	+2, +3	1070	–2.41
Eu	63	151.96	5.259	2.042	0.947	+2, +3	826	–2.40
Gd	64	157.25	7.895	1.802	0.938	+3	1306	–2.40
Tb	65	158.92	8.272	1.782	0.923	+3, +4	1356	–2.39
Dy	66	162.50	8.536	1.773	0.912	+3, +4	1500	–2.35
Ho	67	164.93	8.803	1.766	0.901	+3	1460	–2.32
Er	68	167.26	9.051	1.757	0.890	+4	1497	–2.39
Tm	9	68.93	9.33	1.746	0.880	+2, +3	1545	–2.28
Yb	70	173.04	6.977	1.940	0.860	+2, +3	824	–2.25
Lu	71	174.97	9.842	1.734	0.851	+3	1652	–2.25

*Atomic Radius in Metal

** Standard Electrode Potential (reduction), $M^{3+}3e^- \to M$

However, it is important to observe that the +2 and +4 oxidation states are not so common and are unstable in aqueous solution except for the well known Ce^{4+} salts like ceric sulphate which finds use as an oxidising agent in volumetric analysis.

(iv) It is not possible to explain certain oxidation states, *e.g.*, Sm^{2+} having f^6, Tm^{2+} having f^{13}, Pr^{4+} having f^1 configuration, etc. on the basis of extra stability of an empty, half-filled or complete-filled f subshell. The stabilities of such oxidation states can be explained on the basis of highly complicated thermodynamic and kinetic consideration.

From the above discussion it is evident that although special stability associated with empty, half-filled and completely-filled *f* orbitals has been an important factor, some other factors are also involved in ascertaining the stabilities of the oxidation states.

Compounds with Non-metals and Metalloids

The lanthanides are also known to form well-defined compounds with non-metals and metalloids other than oxygen and halogens. These compounds can metalloids other than oxygen and halogens. These compounds can generally be prepared by the direct combination of the elements at elevated temperatures.

Among these compounds have been Ln_2S_3 (not of Eu), selenide (Ln_2Se_3) and oxy sulphides (Ln_2O_2S). Nitrides (LnN), phosphides (LnP), arsenides (LnAs), antimonides (LnSb) and bismuthides (LnBi) crystallize in the NaCl structure. The carbide (Ln_3C) is known for the lanthanides, Sm–Lu but not for La, Ce, Pr and Nd.

Chemistry of +2 State

It is an anomalous oxidation state. The lanthanides exhibiting this oxidation state can be divided into two categories.

(a) Sm_{62}, Eu_{63} and Yb_{70} : The dipositive ions of these lanthanide (*i.e.*, Sm^{2+}, Eu^{2+} and Yb^{2+}) exist in solution. The standard oxidation potentials at 25°C, in acid solution, of these cations have been given below :

$$Yb^{3+}\,(aq) + e^- \rightleftharpoons Yb^{2+}(aq) \text{-1.15 volts}$$

$$Eu^{3+}(aq) + e^- \rightleftharpoons Eu^{2+}\,(aq) \text{-0.43 volt}$$

$$Yb^{3+}\,(aq) + e^- \rightleftharpoons Yb^{2+}(aq) \text{-1.15 volts}$$

From these values it follows that Sm^{2+},Eu^{2+} and Yb^{2+} ions are strong reducing agents and their reducing strength is in the following order :

$Sm^{2+} > Yb^{2+} > Eu^{2+}$

Sm^{2+} and Yb^{2+} ions are rapidly oxidised by H_3O^+ ions (acidic solution), whereas Eu^{2+} ion is fairly stable and gets only slowly oxidised by H_3O^+ ion.

$$2\ Sm^{2+}\ (or\ Yb^{2+}) + 2H_3O^+ \rightarrow 2Sm^{3+}\ (or\ Yb^{3+}) + 2H_2O + H_2$$

All these cations rapidly get oxidised in presence of oxygen, *e.g.*,

$$4Ln^{2+} + 4H_3O^+\ O_2 \rightarrow 4Ln^{3+} + 6H_2O,$$

where Ln^{2+} can be Sm^{2+} Eu^{2+} or Yb^{2+}.

The compounds of Sm^{2+} Eu^{2+} and Yb^{2+} which have been insoluble in H_2O do not get oxidised by H_2O while hydrated water-soluble compounds of Sm^{2+} and Yb^{2+} get oxidised by their water. Hydrated water-soluble compounds of Eu^{2+} have been more stable.

(b) Ce_{58}, Nd_{60} and Tm_{69} : The compounds having these elements in +2 oxidation state are known only as solid halides. These immediately get oxidised with air.

Of the divalent compound of lanthanides, those of Eu^{2+} ion have been found to be most stable. The compounds of Ln^{2+} ion have been not stable in solution. All the Ln^{2+} compounds are able to decompose water with evolution of H_2.

$$2Ln^{2+} + 2H_2O \rightarrow 2Ln^{3+} + H_2^-$$

The reaction has been, however, sluggish; with Eu the reaction has been so retarded that Eu2+ compounds may be considered as fairly stable in aqueous solutions at an ordinary temperature.

Chemistry of + 4 State

The oxidation state is an anomalous oxidation state. Double salts like $Ce(NO_3)_4.2NH_4NO_3$ and $Ce(SO_4)_2.2(NH_4)_2SO_4.2H_2O$ have also been obtained.

The standard oxidation potentials at 25°C, in acid solution, of Ce^{4+} and Pr^{4+} ions may be put as follow;

$$Ce^{4+} + e^- \rightleftharpoons Ce^{3+}\ +1.74\ volts$$

$$Pr^{4+} + e^- \rightleftharpoons Pr^{3+}\ +2.86\ volts$$

From these values, it is evident that Ce (IV) and Pr (IV) are strong oxidising agents, the latter being by far the stronger of the two. Ce $(SO_4)_2$ generally finds use in volumetric analysis. Ce^{4+} ion is readily reduced to Ce^{3+} ion.

The tetravalent ions of Ce have been found to be stable in the solid state as well as in solution; Pr^{Iv}, Nd^{IV}, Tb^{IV} and Dy^{IV} are stable only in solution.

5. Physical Properties

We have already studied that these elements have similar chemical properties but show some variations in physical properties. Some of these are discussed on next page:

(i) **Ionization Energy :** The values of first ionization energy of lanthanides are about 588 kj mol^{-1} and of the second ionization energy are about 1155 Kj mol^{-1}. These values closely compare with the values of ionization energies (first and second) of the alkaline earth metals particularly with those of calcium.

(ii) **Electronegative Values :** The values of electronegativity of the lanthanides lie in the range 1.0–1.15. The values are of the same order as those of s-block elements. Therefore, lanthanides are expected to form ionic compounds.

(iii) **Metallic Radii :** Generally speaking, metallic radii decrease with increase in atomic number. Same trend is observed in case of lanthanides (except for Eu and Yb). These metallic radii are in the same range as those of s-block (1.74 Å). It suggests that they are fairly large atoms. The large interatomic distance in case of Eu and Yb are because of differences in crystal structure. Eu and Yb have b.c.c. type of structure whereas other members have h.c.p. type of crystal structure.

(iv) **Physical Constants :** Like density, M.P., etc. change gradually except in case of Ce, Eu and Yb. The lower values of Eu and Yb may due to the fact that Yb forms +2 ions which results in slightly weaker metallic bonds, thus M.P. and B.P. are lower.

(v) **Colour :** These metals are silvery white in colour but their ions have colours in both crystalline state and in aqueous solution. The colours of these ions show a definite periodicity which was first pointed out by Main Smith in 1927. The colour of the ion is directly connected to the number of unpaired electrons in 4f-subshell. The colour of the elements with f electrons is the same as that of elements with (14–X) f electrons. This is illustrated in Table 6.

Absorption spectra of the compounds of Ln^{3+} ions in these case show a very sharp line band in u.v. visible or near i.r. region. These bands are due to internal transitions in partly filled 4*f*-shell electron and are called *f*–*f* transitions.

In case of *d–d* transitions broad bands are observed (compare with *f–f* these elements. Because 4*f*-orbitals are shielded from external perturbation so the complexing agents have little effect on the spectral bands of these lanthanide ions.

Table 6 : Colour of Ions 4*f*- Elements

Elements	*At. No.*	*Unpaired f-elements*	*Colour of M^{3+} ion*
La	57	0	Colourless
Ce	68	1	Colourless
Pr	69	2	Yellow-green
Nd	60	3	Red-Violet
Pm	61	4	Pink-yellow
Sm	62	5	Yellow
Eu	63	6	Pale pink
Gd	64	7	Colourless
Tb	65	6	Pale pink
Dy	66	5	Yellow
Ho	67	4	Pink-yellow
Er	68	3	Pink
Tm	69	2	Pale-green
Yb	70	1	Colourless
Lu	71	0	Colourless

The following points are worth-nothing from the Table 1.4

(i) As the colour of the two lanthanides having same number of 4f unpaired electrons is same, it is believed that the colour of the ions is dependent on the number of unpaired f electrons.

(ii) The lanthanide cations having x electron in 4f-orbitals are having the same colour as the lanthanide cations having (14–x) electrons in the 4f orbital. Thus Nd^3 (x = 3) and Er^{3+} (14–3 = 11) cations having 3 and 11 electrons in the 4f orbital respectively are having the same colour (reddish). This reveals that 4f orbitals are the source of the colour of lanthanide ions.

(iii) Lanthanide ions containing empty, half-filled or completely filled 4f orbital are colourless, *e.g.*, La^{3+} ($4f^0$), Gd^{3+}($4f^7$) and Lu^{3+} ($4f^{14}$) ions are colourless. However, the relation between colour and number of 4f unpaired electron each.

(a) Ce^{3+} and Yb^{3+} are colourless although both contain one unpaired electrons fails to explain the following points.

(b) Dipositive and tetrapositive ions which are isoelectronic with tripositive ions (ions having the same number of unpaired electrons) have different colours.

The discrepancy is attributed to the fact that colour is derived from absorption of certain wavelength in visible region. The absorption in turn is dependent on the energy required by electrons to get excited from one level to another. The energy in turn is dependent upon the number of available orbitals to which 4*f* electrons can get excited in the individual.

Origin of Colour

The colours are because of *Laporte-forbidden f–f transitions, i.e., transitions* between the J states of 4fn configuration. Hence the absorption bands of Ln^{3+} ions (except Ce^{3+} and Yb^{3+} ions) are *very weak* but *sharp* when compared to those of *d*-block elements. Many of these bands have been found to be *line-like* and *become even narrower as the temperature is lowered.* These narrow bands appear due to *f–f* transition and are independent of the nature of the anion present. As 4*f*-electrons lie deep inside the atom, the colours of Ln^{3+} ions do not get effected by changing the anion.

The colour one sees is the result of absorption of light of certain wavelengths and transmission of light of other wavelengths. The coloured ions absorb in the visible region, and on occasion in the ultraviolet region.

The colourless ions absorb either in the ultraviolet (Ce^{3+}, Gd^{3+} ions) or in the infrared region (Yb^{3+} ion). The Ln^{2+} ions absorb strongly in the ultraviolet. The only Ln^{4+} ion stable in aqueous solution, the Ce^{4+} ion, is known to absorb in the blue ultraviolet regions.

Laporte-permitted bands because of the transitions of $4f^n \rightarrow 5d^1$ type have been observed in Ce^{3+}, Tb^{3+}, Sm^{2+}, Eu^{2+} and Yb^{2+}. These bands are *strong and broader,* because the transition is considerably influenced by the chemical environment.

A charge transfer phenomenon is also observed in certain lanthanide ions, *e.g.*, the orange-red colour of Ce^{4+} is because of the *electronic transition from the ligand orbital to the f-orbital of cerium.* Compounds of Eu^{3+} with reducing anions are yellow because of *electron transfer from the metal.*

Magnetic Properties

Due to the presence of unpaired electron in 4f-orbitals, all the lanthanides ions except those of La^{3+}, L^{u2+}, Yb^{3+} and Cu^{3+}, show paramagnetic behaviour. The magnetic moments of those ions do not obey the "spin only" formula $\mu = [n\ (n+2)]$. This formula was able to explain the magnetic moments to transition elements. n in this formula represents the number of unpaired electrons presets in the ion.

According to modern view paramagnetism is contributed by spin of the electron as well as by its orbital motion. In *d*-type transition elements, d-orbitals are not well shielded and they can participate in bond formation. In such a case magnetic moment is due to electron spin only and so depends mostly on the number of unpaired electrons. In case of lanthanides 4*f*-electrons are well shielded and can not participate in bond formation so they are well shielded from one quenching effect of the environments so the magnetic moments of lanthanides are calculated by taking into consideration spin and orbital contributions, a more complex formula is used

$$\mu = \sqrt{[4S(S+1) + (L+1)]}$$

where L is the orbital quantum number and S is the spin quantum number. The results calculated using this formula are found to be in agreement with experimental values. La^{3+} is diamagnetic (due to f^0) the values of magnetic moments increase from La to neodymium, which has maximum value. Then a decrease is observed for Sm ($\mu = 1.47$). Magnetic moment values again increase and dysprosium and holmium and holmium have maximum values. These again fall and reach Lu which is diamagnetic (due to f^{14}).

Complexes

Although, the lanthanide ions are having a high charge (+3), their large size (0.85– 1.03 A°) imparts them *low charge density* (charge to size ratio) with the result they can not bring about much polarisation nd hence are not having much tendency to form complexes. Their complexes with unidentate ligands are very few. However, complexes with a few chelating ligands such as β-diketone, oximes and ethylenediamine tetra (EDTA) are fairly common.

The following points are of interest regarding complex formation.

(i) The tendency to form complexes and their stability tends to increase with increasing atomic number. This property finds use in the separation of lanthanides from one another. However, this order becomes reverse in case of hydrated ions, Ln^{3+} (aq) *i.e.,* the tendency of complex formation among the hydrated lanthanide ions gets decreased with increasing atomic number.

(ii) with a specific ligand, the order of complex formation for the Ln^{2+}, Ln^{3+} and Ln^{4+} ions has been as follows:

$$Ln^{4+} > Ln^{3+} > Ln^{2+}$$

Reactivity of Lanthanides

As all the lanthanides are having similar outer electronic configuration and display mainly +3 oxidation state in their compounds, they show exceedingly similar chemical properties. Their similarity has been somewhat more than that among the ordinary transition elements because lanthanides differ mainly in the number of 4*f* electrons which are very effectively shielded from interaction with other elements by the overlapping of 5*s*, 5*p* and 6s electrons.

Moreover, because of lanthanide contraction there occurs very small difference (maximum 0.21 Å) among the size of the trivalent lanthanide cations. Thus, for practical purposes the size of these ions is almost identical which gives rise to similar chemical properties of the elements.

The lanthanides have been highly reactive which has been in agreement with the values of their ionisation energy and electronegativity. The ionisation energies of lanthanides have been found to be somewhat comparable with those of alkaline earth metals particularly calcium. Hence like alkaline earth metals, lanthanides are highly electropositive and very reactive metals which are evident from the following points.

(i) Although, they are silvery white metals, they get tarnished readily on exposure to air.

(ii) All of them burn in air yielding the sesquioxide, Ln_2O_2 except cerium which gives CeO_2. Ytterbium resists the action of air even at 1000°C because of the formation of a protective coating of its oxide.

(iii) They dissolve slowly in cold water but more rapidly in the warm water liberating hydrogen.

(iv) They react with hydrogen forming non-stoichiometric hydrides approaching LnH_2 and LnH_3 in composition.

(v) They react with nitrogen (especially when warmed) to form corresponding nitride, LnN.

(vi) Lanthanides also react with non-metals such as halogens, sulphur, phosphorus, carbon and silicon to form corresponding compounds.

(vii) Their high oxidation potentials reveal their strong electropositive character. Thus, they act as strong reducing agents.

(viii) All get attacked by acids liberating hydrogen.

3

Classification of d-block Elements

INTRODUCTION

It is significant to not that the compounds in the lower oxidation state *i.e.*, 2+ are reducing as in the case of Cr^{2+}, Fe^{2+}, etc. The oxides are also reducing and basic in character.

In the 3+ oxidation the oxides are amphoteric, *e.g.*, Cr_2O_3 with acids forms chromic salts while with alkalies it forms chromites. Same is the case with MnO_2 containing manganese in the 4+ oxidation state.

In the higher oxidation states *i.e.*, 6 +, 7+ the oxide are acidic. *e.g.*, CrO_3, MnO_3,FeO_3 and Mn_2O_7 and also oxidizing.

The Table 1.7 showing the different oxidation states and the electronic configuration tends to show some interesting generalizations, *e.g.*,

(i) $3d^5$ configuration is associated with the stability of Mn^{2+} and Fe^{3+}, $3d^{10}$ with that of Cu^+.

(ii) 3d electronic configuration in the case of Sc^{3+} , Ti^{4+}, V^{5+} is associated with the colourless and diamagnetic ions as in the case of scandium, TiO_2 and $TiCl_4$ and vanadates Vo_4^{3-} etc. Similarly, $3d^{10}$ electronic configuration explains the formation of colourless CU and Zn^{2+}.

(iii) In case of elements from Sc (21) to Mn (25) the group oxidation states are observed upto attainment of d^5 electronic configuration *i.e.*, half-filled d orbital *e.g.*, Sc^{3+} , Ti^{4+} ,V^{5+}, Cr^{6+} and Mn^{7+}.

However, from Fe (26) to *i.e.,* from $3d^6$ to $3d^{10}$ electronic configuration the group oxidation is not attained.

Following points are observed in the oxidation states of transition elements :

(i) Many elements exhibit oxidation states which are different from each other only by one unit. This is in marked contrast to the s-block elements which are known to exhibit only the group valency and the p-block elements which are known to exhibit either the maximum group valency or valency less by two units. *e.g.,* Sn and Pb (group IV A elements) exhibit 2 and 4 valencies.

(ii) In case of the first elements of the first series *i.e.,* upto Mn in which 3d subshell is having only unpaired electrons, the minimum oxidation state would be given by the number of electrons in the outer s-subshell and the maximum oxidation state would be given by the sum of the outer s-and d-electrons.

In the next five elements in which 3d subshell is having some paired electrons, the minimum oxidation state would be still equal to the number of the outer s-electrons, whereas the maximum oxidation state is not related at all with the electronic configuration ; in fact it has been rarely higher than +2 or +3.

(iii) Across the period, *i.e.,* from left to right, the (+2) state becomes more stable while the (+3) state becomes less stable. Very high oxidation states cannot exist as the free ions, although oxyions do exist.

As the electronic structures of the atoms of the second and third transition do not exactly follow the pattern of the first series, the above points would be only partially true in these two series.

PROPERTIES OF BERYLLIUM PHYSICAL

1. It is a silvery white (or steel grey) metal.
2. It is malleable, easily forged and takes a high polish.
3. Commercial beryllium is so hard that it will scratch glass and is also very brittie.
4. It is a light metal (SP. g. 1.83) and it melts at 1278°C, the metal is volatile.

5. The sp. heat, thermal conductivity and specific resistance of beryllium increase with increase in temperature.

Beryllium has high transparency towards X-rays, and this is distinguished from most of the metals. Two isotopes of beryllium 8 and 9 have been reported.

Chemical Properties

1. In its chemical nature beryllium resembles aluminium and magnesium.
2. The metal is oxidised in the moist air.
3. When heated to redness in oxygen it forms a protective layer of oxide round the metal.

$$2Be + O_2 \rightarrow 2BeO$$

Powdered metal burns brilliantly in air forming BeO.

4. *Reaction with H_2O :* Water or steam is not decomposed by beryllium even at red heat.
5. *Reaction with halogens :* The metal directly combines with halogens.

$$Be + Cl \rightarrow BeCl_2$$

$$Be + Br_2 \rightarrow BeBr_2$$

It unites with idodine with difficulty, but with incandescence with chlorine and bromine.

Reaction with non-metals : It does not directly combine with sulphur, carbon and boron, but unites indirectly with boron and carbon.

7. *Reaction of acids :* Dilute acids like H_2SO_4, HCl dissolve it with the evolution of hydrogen.

$$Be + H_2SO_4 \rightarrow BeSO_4 + H_2$$

Be reacts with conc. sulphuric acid with the evolution of sulphur dioxide.

$$Be + 2H_2SO_4 \rightarrow BeSO_4 + SO_2 + 2H_2O$$

Conc. HNO_3 has hardly any action. At higher dilutions beryllium nitrate is formed and No is evolved with dil. HNO_3.

8. *Reactions of alkalies :* Ammonia does not attack it, but caustic alkali acts on it similar to aluminium.

$$Be + 2NaOH \rightarrow Be\,(ONa)_2 + H_2$$

9. *Other reactions :* Beryllium decomposes partially SiO_2 into si.

$$2Be + SiO_2 \rightarrow 2BeO + Si$$

Nitrogen : Beryllium combines with N_2 above 900°C.

With metals : Beryllium forms alloys with various metals. Alloys of Cu and Be are prepared by heating a mixture of their oxides with carbon in an electric furnace.

Uses

1. Be is used as windows in X-rays tubes. Its transparency for X-rays is 17 times greater than that of Al.
2. Beryllium is used in the production of alloys.
   (a) Be –Fe alloy is used in the manufacture of aeroplane parts.
   (b) Be–Cu ball bearing are used in place where magnetic effect must be reduced to minimum.

   Be–Cu watches in market are unaffected by a magnetic field.

   (c) Alloys of Be with Al.Ni and Co are used in aircraft construction.
3. Beryllium oxide is used in the manufacture of synthetic gems, abrasives and dental products.
4. Beryllium is used for making electrodes of neon-signs because of its high melting points.
5. Beryllium finds application in fields of developing rockets, missiles and high speed aircrafts.
6. Beryllium oxide finds use in gas. mantles along with beryllium nitrate.
7. Beryllium compounds are used in the "coating" of some form of fluorescent lamps.
8. Beryllium has recently been used as an igredient of a new textile of a new textile fibre of the rayon type made of seaweeds.

COMPOUNDS OF BERYLLIUM OXIDE, BEO

Preparation

1. Beryllium oxide also known as beryllia is obtained by the ignition of basic carbonate, $BeCO_3.3Be(OH)_2.2H_2O$.

$$BeCO_3 \rightarrow BeO + CO_2$$

2. By heating beryllium nitrate.

$$2Be(NO_3)_2 \rightarrow 2BeO + 4NO_2 + O_2$$

3. By heating the element strongly in air or oxygen.

$$2Be + O_2 \rightarrow 2BeO$$

Properties

1. It is a white amorphous power, insoluble in water. It melts at 2450°C.
2. It dissolves in HCl or HNO_3, more acted upon by conc. H_2SO_4.

$$BeO + H_2SO_4 \rightarrow BeSO_4 + H_2O$$

3. HF converts it into fluoride, BeF_2.
4. In high vacuum, beryllium emits a blue fluorescence.
5. It is basic but feebly acidic.
6. When fused in an electric furnace it forms a crystalline mass which is harder than diamond.
7. Alkalies dissolve it readily.

$$BeO + NaOH \rightarrow HO.\ BeONa$$

Uses

1. It is used in the manufacture of incandescent gas mantles.
2. It is used as an abrasive.

BASIC BERYLLIUM ACETATE, 3BE$(CH_3COO)_2$. BEO

Preparation

1. **Urbain method :** The mixture of beryllium hydroxide and dilute acetic acid is evaporated to dryness. The residue obtained in boiling anhydrous acetic acid and cooled. On extracting with chloroform, the basic beryllium acetate dissolves.

$$3Be(OH)_2 + 3(CH_3COO)_2Be \rightarrow 3Be(CH_3COO)_2.BeO + 4H_2O$$

2. **Parson's methods :** By heating a mixture of beryllium carbonate with glacial acetic acid.

$$6BeCO_3 + 6CH_3COOH \rightarrow 3Be(CH_3COO)_2.BeO + 6H_2O + 6CO_2$$

Properties

Basic beryllium acetate is a acetate is a volatile crystalline solid. It is insoluble in water but dissolves in lower alcohols and other non-polar organic solvents. It is obtained as octahedral crystals when its solution in organic solvents is crystallised. The compound melts at 283°C and boils at 330°C without decomposition.

Structure

From the properties of the basic beryllium acetate it follows that compound is cavalent. The structure has been studied by X-ray methods. It has been shown that the molecule has a terahedral symmetry. As shown in the structure the central oxygen along is tetrahedrally surrounded by four beryllium atoms in such a way that the four acetate groups form the edges of the tetrahedron. The studies of *Pauling and Sherman (1934)* have shown that in the crystal cell there are four molecules of basic acetate and the arrangement corresponds with diamond structure.

Beryllium Hydroxide, $Be(OH)_2$

It is prepared as a white gelatinous precipitate by adding NH4OH to the solution of beryllium salts.

$$BeSO_4 + 2NH_4OH \rightarrow Be(OH)_2 + (NH_4)SO_4$$

Properties

It is white gelatinous when precipitated and dried mass is amorphous substance. It is soluble in excess of alkali and acids. Freshly precipitated hydroxide dissolves in the solution of ammonium carbonate. It is also four insoluble in ethyl amine. When $Be(OH)_2$ is heated strongly, it changes to oxide.

Beryllium Fluoride, BeF_2

It is obtained by heating the double salt, $BeF_2.2NH_4F$ in the absence of air. It is a hygroscopic substance easily soluble in water. Like BeO, BeF_2 is also covalent in nature.

Beryllium Chloride, $BeCl_2$

It is prepared by the action of (a) chlorine or (b) HCl on metal.

Anhydrous beryllium chloride is obtained by passing Cl2 over a mixture of BeO and carbon.

$$BeO + C + Cl_2 \rightarrow Becl_2 + CO$$

Properties

It is a white solid which melts at 405°C. It dissolves in water with evolution of heat.

The aqueous solution of $BeCl_2$ is acidic due to partial hydrolysis. When aqueous solution is evaporated in presence of HCl gas, crystals of the composition of $BeCl_2.4H_2O$ are obtained.

Anhydrous $BeCl_2$ shows behaviour of covalent compound. It dissolves in organic solvents like alcohol, pyridine, ether, etc. It forms addition compounds with ether, $BeCl_2$. $(C_2H_5)_2O$.

Beryllium halides are coordinatively unsaturated and exist is gas phase. However, in condensed phase they achieve coordination number of four through polymerization (Fig. 1.11) and formation of compounds like $BeCl_2.(OC_2H_5)_2$ or an ionic species like $[BeF_4]^{2-}$.

Beryllium Sulphate, $BeSO_4$

It is obtained by the action of hot concentrated sulphuric acid on beryllium oxide. On cooling, the crystals of beryllium sulphate, $BeSO_4$. $4H_2O$ are obtained. It hydrolyses aqueous solution to form the hydroxide.

$$BeSO_4 + 2H_2O \rightarrow Be(OH)_2 + H_2SO_4$$

Beryllium nitrate : It maybe prepared (i) By double decomposition of $BeSO_4$ and $Ba(NO_3)_2$ (ii) By dissolving $Be(OH)_2$ in nitric acid (iii) By treating beryllium carbonate with nitric acid.

$$BeSO_4 + Ba(NO3)_2 \rightarrow Be(NO_3)_2 + BaSO_4$$

$$Be(OH)_2 + 2HNO_3 \rightarrow Be(NO_3)_2 + 2H_2O$$

$$BeCO_3 + 2HNO_3 \rightarrow Be(NO_3)_2 + H_2O + CO_2.$$

It is a colourless crystalline substance. When heated at about 90°C it forms beryllium oxide. It is used in gas mantle industry as hardening agent.

GALLIUM

Discovery : The discovery of indium was made earlier than gallium due to which there existed a gap between aluminium and indium. The discovery of gallium by Boisbaudran in 1875 filled this gap. After streneous research work since 1863 so as to discover this element in zinc blende with the help of spectroscope-Gallium in honor of his native country Gallia (Latin name for France).

Occurrence

Gallium occurs widely distributed in nature in minute quantities. Traces of gallium, and large amounts of lead occur in almost all samples of zinc blende, and if zinc is purified by fractional distillation the lead residue may be having as much as 0.1 per cent of gallium. This is probably the best source of the element. Gallium is often present in Kaolin (China clay). Germanite is the mineral richest in gallium and sometimes it is having as much as 1.8 per cent of the metal. The recent discovery of gallium oxide (Ga_2O_3) in the ash of coal from the Northumbrian coal fields, reveals another new source of the element.

Extraction

1. From Germanite

When powdered mineral is distilled with concentrated hydrochloric acid in a stream of chlorine, the germanium is volatilized as the chloride, and when the filtered residue is made to treat with ammonia, a mixture of oxides and hydroxides, including gallium hydroxide is thrown down. The precipitate is dissolved in hydrochloric acid and copper is removed as sulphide and the gallium then precipitated as potassium gallium ferrocyanide. This when fused with caustic soda yields a water soluble sodium gallate. Now the metal is obtained by the electrolysis of solution of sodium gallate.

2. From zinc Blende

Zinc blende having gallium is dissolved in aqua regia. Solution is filtered and then treated with metallic zinc which precipitates metals like Cu, Pb, Cd, Hg and Ag. These are filtered and solution is boiled once again with excess of metal zinc. A white precipitate having basic salt of zinc, gallium, iron and chromium and also alumina and silica is formed. This precipitate obtained is dissolved in hydrochloric acid and hydrogen sulphide is passed. Precipitate obtained is filtered off and the filtrate is neutralized with ammonia and then a buffer solution having ammonium acetate and acetic acid is added. Hydrogen sulphide is again passed, which precipitates zinc and gallium sulphides together.

This precipitate of sulphides is washed with water and dissolved in hydrochloride acid when chlorides of zinc and gallium are obtained. To the cold solution of this chloride mixture is added a sodium carbonate solution. Gallium gets precipitated out first and basic zinc carbonate precipitated there after.

The precipitate, mostly having basic gallium carbonate is dissolved in sulphuric acid and the solution gently heated until excess of the acid is removed. The residue is dissolved in excess of water and boiled for few hours when basic gallium sulphate gets precipitated. The precipitate is washed and the process is repeated in order to get pure sulphate.

Gallium sulphate precipitate is now made to dissolve in sulphuric acid and then treated with excess of caustic potash. A soluble potassium gallate is probably formed which can be electrolysed using a large platinum cathode.

Zn-Ga blende $\xrightarrow{\text{aqua regin}}$ Solution and Ag ppted $\xrightarrow{\text{Zn}}$ Cu, Cd, Pb, Hg Filtrate $\xrightarrow{\text{Boiled with Zn}}$ White ppt of basic salts of Zn, Ga, Fe, Cr and alumina silica

$\downarrow$ HCl

Solution of $ZnCl_2$, $GaCl_3$ $\xleftarrow{\text{HCl}}$ Ppt. of ZnS and Ga_2S_2 $\xleftarrow{H_2S}$ Solution $\xleftarrow{NH_4OH}$ Filtrate $\xleftarrow{H_2S}$ Solution

$\downarrow$ Na_2CO_3

First Ppt. of Ga basic carbonate

$\downarrow$ H_2SO_4

Soluble $Ga_2(SO_4)_3$

$\downarrow$ Excess water and boiled

White pp. of basic gallium sulphate $\xrightarrow[\text{treated with excess KOH}]{\text{Dissolved in } H_2SO_4 \text{ and}}$ Solution $\xrightarrow[\text{Pt-Cathode}]{\text{Electrolysed}}$ Ga metal

Metallurgy

It is possible to obtain gallium in the pure form by the electrolysis of an alkaline solution of gallium salt using platinum cathode. The metal gets deposited on the platinum cathode as a liquid and sometimes in a solid state as a tree. The trees formed on the cathode would be hard and fairly stable provided the temperature of the bath is kept 10° below the melting point.

Properties

Physical : Gallium occurs as a grayish metal resembling steel but in a molten state it appears like tin or silver. It can be obtained in the form of octahedral crystals on dipping a platinum wire supporting a little solid gallium into molten metal cooled to 10-15°C below its melting point. The solid metal is having a crystalline structure. The metal is hard and leaves a bluish grey mark on the paper. The melting point of the metal is 30.15°C and boiling point 1700°C. Gallium adheres to the glass surface on which it forms a fine mirror.

Chemical-Action of Air : Gallium is not effected in air and maintains its lusture. On heating strongly, the metal, however, oxidizes only superficially.

Action of water : Air free water does not have any effect on the metal, but in presence of air, it gradually gets tarnished.

Action of Acids : Cold or hot hydrochloric acid is able to dissolve the metal with the evolution of hydrogen. In nitric acid also it is soluble but the best solvent is aqua-regia.

$$7Ga + 3HCl \rightarrow GaCl_3 + H_2 \uparrow$$

Action of Alkalies : Hydrogen is liberated if gallium is treated with a solution of caustic potash.

$$2Ga + 6KOH \rightarrow 2K_3HaO_3 + 3H_2 \uparrow$$

Uses

Although gallium is a costly metal, yet it finds some good applications.

1. Due to the wide range of temperature between is melting point and boiling point, it finds use in quartz thermometers for

measuring high temperatures, (above 700°C) far above the mercury thermometers.

2. Gallium being non-poisonous can be used in teeth filling.
3. It also finds use in lighting tubes, photoelectric cells, infra red transmitters and electron tubes.
4. A certain alloy of gallium aluminium finds use in quartz vacuum lamps and gallium-cadmium alloy finds use in monochromatic lamps.

Gallium Compounds

Oxide of Gallium : There are two oxides of gallium, gallium sesquioxide or gallic oxide, Ga2O3 and gallous oxide, GaO. Gallic oxide can be prepared by heating gallium nitrate. This oxide on heating in a stream of dry hydrogen, gets partially sublimed and partly reduced into a bluish-grey gallous-oxide, GaO. Boisbaudraun obtained a brown gallous oxide by the action of water on gallous chloride.

$$GaCl_2 + 2H_2O \rightarrow Ga(OH)_2 + 2HCl$$

$$Ga(OH)_2 \rightarrow GaO + H_2O$$

Gallous oxide is colourless. It exists in dimorphic forms with a transition temperature of about 500°C. Gallic Ga2O3 is polymorphic like alumina. Gallic oxide is infusible at a red heat. The oxide is soluble in mineral acids, alkali hydroxide and ammonia solution, but the ignited oxide is insoluble in acid or alkali. Hydrogen is able to reduce both gallic and gallous oxides to metal gallium.

Electronegativities

The evaluation of electronegativities across a series of transition metal element is complicated by oxidation states. It is possible that element participating directly in bonding may have a somewhat greater screening effect than shared electrons. Thus, in lower oxidation states the electronegativity may be lower. However, this effect may not be very large. The electronegativities increase slowly as the number of d-electrons increase in a transition series. In a group, the electronegativities go down generally. The range from low to high is from La (57), only a little higher than Ba (56), Cu (29), about the same as Cd (48).

Electronegativities of transition elements (Pauling Scale) are as follows:

Table 1

Sc	*Ti*	*V*	*Cr*	*Mn*	*Fe*	*Co*	*Ni*	*Cu*	*Zn*
1.3	1.5	1.6	1.6	1.5	1.8	1.8	1.8	1.9	1.6
Y	Zr	Nb	Mo	Tc	Ru	Rh	Pd	Ag	cd
1.2	1.4	1.6	1.8	2.2	1.9	2.2	2.2	1.9	1.7
La	Hf	Ta	W	Re	Os	Ir	Pt	Au	Hg
1.1	1.3	1.5	1.7	1.9	2.2	2.2	2.2	2.4	1.9

Electronic Configuration

In the various transition elements of a series of the configuration of the outer most shell remains unchanged while one electron is added continuously until its full capacity (*i.e.,* 10) to the (n–1) d orbitals. On the basis of the number of (n–1) d orbitals it is possible to divide the whole of the transition elements into three series, viz when (n–1) d orbitals are 3, 4 and 5. The three series, are respectively termed as 3d (first), 4d (second) and 5d (third) series. Following points are to be remembered while going through the configuration of the various transition elements.

(i) We know that the orbitals are filled up in order of their increasing energy, *i.e.,* an orbital of lower energy get filled first followed by orbital of higher energy. Thus, 4s orbital having lesser energy is filled first to its capacity ($4s^2$) than the 3d orbital having higher energy. Thus, when 3p orbitals are filled to their capacity as in Kr-atomic number 18; the 19th and 20th electrons in k and Ca respectively enter 4s orbital rather tan 3d orbital the 3d orbital starts filling at scandium (atomic number 21) only when 4s orbital becomes full to its capacity.

Table 2

Element	*Symbol*	*At. No.*	*Electronic configuration*
Krypton	Kr	18	Is^2, $2s^2p^6$, $3s^2p^6$
Potassium	K	19	Is^2, $2s^2p^6$, $3s^2p^6$,$4s^1$
Calcium	Ca	20	Is^2, $2s^2p^6$, $3s^2p^6$,$4s^2$
Scandium	Sc	21	Is^2, $2s^2p^6$, $3s^2p^6d^14s^2$

(ii) We know that the exactly half filled and completely filled d-orbitals are extra stable. This explains the electronic configuration of chromium (at. No. 24) as

$$[Ar]3d^5.4s^1 \text{ but not as } [Ar]3d^1,4s^2$$

The electronic configurations of elements of the first series can be put as follows :

Table 3 : Electronic Configuration of Elements of the First Series

Element	*Symbol*	*At.No.*	*Electronic configuration*
Scandium	Sc	21	$1s^2, 2s^2p^6, 3s^2p^6d^1,4s^2$
Titanium	Ti	22	$1s^2, 2s^2p^6, 3s^2p^6d^2,4s^2$
Vanadium	V	23	$1s^2, 2s^2p^6, 3s^2p^6d^3,4s^2$
*Chromium	Cr	24	$1s^2, 2s^2p^6, 3s^2p^6d^5,4s^1$
Manganese	Mn	25	$1s^2, 2s^2p^6, 3s^2p^6d^5,4s^2$
Iron	Fe	26	$1s^2, 2s^2p^6, 3s^2p^6d^6,4s^2$
Cobalt	Co	27	$1s^2, 2s^2p^6, 3s^2p^6d^7,4s^2$
NIckel	Ni	28	$1s^2, 2s^2p^6, 3s^2p^6d^8,4s^2$
*Copper	Cu	29	$1s^2, 2s^2p^6, 3s^2p^6d^{10},4s^2$
Zinc	Zn	30	$1s^2, 2s^2p^6, 3s^2p^6d^{10},4s^2$

***Note :** The anomalous electronic configuration of Cr and Cu arises because of the extra stability of the exactly half-filled and completely filled d-orbitals.

It is quite difficult to predict the electronic configurations of elements of second and third series. Many other interactions such as electron-electron repulsion and nuclear-electron attraction play an important role in ascertaining the configuration of some of the members. The electronic configurations of the elements of these two series are given as follow:

The third transition series begins at lanthanum (La, Z =57) in which the additional electron starts entering 5d subshell. However, after La (Z = 57), the next electron fails to enter the 5d subshell but it starts entering the 4f subshell as in cerium (Z =58) and the 5d subshell gets second electron only when the 4f subshell becomes full to its capacity, *i.e.,* 14 which is achieved in lutetium (Lu Z = 71.

Table 4 : Electronic configuration of elements of the second series.

Element	*Symbol*	*At.No.*	*Electronic configuration*
Yttrium	Y	39	$1s^2, 2s^2p^6, 3s^2p^6d^{10}, 4s^2p^6d^1 5s^2$
Zirconium	Zr	40	$1s^2, 2s^2p^6, 3s^2p^6d^{10}, 4s^2p^6d^2 5s^2$
**Niobium	Nb	41	$1s^2, 2s^2p^6, 3s^2p^6d^{10}, 4s^2p^6d^4 5s^1$
*Molybdenum	Mo	42	$1s^2, 2s^2p^6, 3s^2p^6d^{10}, 4s^2p^6d^5 5s^1$
Technetium	Tc	43	$1s^2, 2s^2p^6, 3s^2p^6d^{10}, 4s^2p^6d^5 5s^2$
**Ruthenium	Ru	44	$1s^2, 2s^2p^6, 3s^2p^6d^{10}, 4s^2p^6d^7 5s^1$
**Rhodium	Rh	45	$1s^2, 2s^2p^6, 3s^2p^6d^{10}, 4s^2p^6d^8 5s^1$
**Palladium	Pd	46	$1s^2, 2s^2p^6, 3s^2p^6d^{10}, 4s^2p^6d^{10} 5s^0$
*Silver	Ag	47	$1s^2, 2s^2p^6, 3s^2p^6d^{10}, 4s^2p^6d^{10} 5s^1$
Cadmium	Cd	48	$1s^2, 2s^2p^6, 3s^2p^6d^{10}, 4s^2p^6d^{10} 5s^2$

Note : *The irregurlarity in the configuration of Mo($4d^5, 5s^1$) and Ag ($4d^{10}, 5s^1$) occurs because of higher stability of exactly half-filled and completely filled subshells.

** It is possible to explain the anomalous configuration of Nb, Ru, Rh and Pd on the basis of the nuclear electron and electron-electron forces.

Table 5 : Electronic configuration of the third series.

Element	*Symbol*	*At.No.*	*Electronic configuration*
Lanthanum	La	57	$2, 8, 18, 4s^2p^6d^{10}f^0, 5s^2p^6d^1, 6s^2$
Hafnium	Hf	72	$2, 8, 18, 4s^2p^6d^{10}f^{14}, 5s^2p^6d^2, 6s^2$
Tantalum	Ta	73	$2, 8, 18, 4s^2p^6d^{10}f^{14}, 5s^2p^6d^3, 6s^2$
Tungsten	W	74	$2, 8, 18, 4s^2p^6d^{10}f^{14}, 5s^2p^6d^4, 6s^2$
Rhenium	Rh	75	$2, 8, 18, 4s^2p^6d^{10}f^{14}, 5s^2p^6d^5, 6s^2$
Osmium	Os	76	$2, 8, 18, 4s^2p^6d^{10}f^{14}, 5s^2p^6d^6, 6s^2$
Iridium	Ir	77	$2, 8, 18, 4s^2p^6d^{10}f^{14}, 5s^2p^6d^7, 6s^2$
Platinum	Pt	78	$2, 8, 18, 4s^2p^6d^{10}f^{14}, 5s^2p^6d^{10}, 6s^0$
Gold	Au	79	$2, 8, 18, 4s^2p^6d^{10}f^{14}, 5s^2p^6d^{10}, 6s^1$
Mercury	Hg	80	$2, 8, 18, 4s^2p^6d^{10}f^{14}, 5s^2p^6d^{10}, 6s^2$

These fourteen elements from Ce (Z = 58) to Lu (Z = 71) are therefore termed as *f-block elements;* and since these are kept within the transition series, these are also termed as *inner transition elements; further since they* are kept after lanthanum these are also termed as *lanthanides.* These fourteen elements are having similar configuration in the outer shells, hence they are very similar in their physical and chemical properties and discussed in detail in a separate chapter.

Variable Oxidation States

This has been one of the most striking features of the transition elements. All transition elements, except the first and the last members of each series show variable oxidation states. The ability of the transition elements to form multiple oxidation states has been attributed to the fact that the ns as well as (n–1)d electrons are of nearly the same energy level. Hence, both ns and (n–1) d electrons are available for bond formation. Thus, the variable oxidation states of transition elements are related to their electronic configuration which is evident from the following table to the oxidation state of elements of the first transition series.

The filling up of the d orbitals is not quite regular. Due to less energy difference between the 3d and 4s orbitals chromium and copper are respective electronic configuration states as $3d^5$ 4s1 and $3d^{10}$ $4s^1$ and not as $3d^4$ $4s^2$ and $3d^9$ $4s^2$ respectively as expected.

Table 6 : Relation Between outer Electronic Configuration and Oxidation States of the Elements of the First transition Series

Element	*Outer elec. configuration*		*Oxidation states*
Sc	$3d^1$	$4s^2$	+2, +3
Ti	$3d^2$	$4s^2$	+2, +3, +4
V	$3d^3$	$4s^2$	+2, +3, +4, +5
Cr	$3d^5$	$4s^1$	+1, +2, +3, +4, +5, +6
Mn	$3d^5$	$4s^2$	+2, +3, +4, +5, +6, +7
Fe	$3d^6$	$4s^2$	+2, +3, +4, +5, +6
Co	$3d^7$	$4s^2$	+2, +3, +4
Ni	$3d^8$	$4s^2$	+2, +3, +4
Cu	$3d^{10}$	$4s^1$	+1, +2
Zn	$3d^{10}$	$4s^2$	+2

The cause of the variable oxidation states can also be attributed to the fact that the energy of oxidation has been sufficient to raise the 3d electrons to the 4p orbitals thus making electrons available for bond formation.

Hybridization of the valency electrons among the various available orbitals takes place ensuring equivalence.

A close look at Table 6 reveals and explains some interesting facts:

(i) Scandium is 3+ involving $3d^14s^2$ electrons.

(ii) Titanium is having $3d^24s^2$ as the electronic configuration in the valence shells. Thus $4s^2$ electrons are associated with dipositive character. $4s^2$ and $3d^1$ electrons contribute to its tripositive character while $4s^2$ and $3d^2$ electrons explain the tetrapositive character.

(iii) In the case of vanadium, oxidation states varying between 2+ to 5+ are obtained. Again 2+ character could be explained by the participation of $4s^2$ in the bond formation and the remaining oxidation states *i.e.*, 3+, 4+, 5+ due to 1, 2 and 3 electrons of 3d orbital respectively.

(iv) Chromium exhibits 2+, 3+ and 6+ oxidation states. The compounds formed in 2+ oxidation state are known as chromous compounds. In the 3+ oxidation state the compounds are known as chromic ones like Cr_2O_3, etc. In the 6+ oxidation state the compounds like chromium trioxide, chromates and dichromates are obtained. In fact dichromates are formed by the condensation of CrO_4^{2-}.

$$2CrO_4^{2-} + 2H^+ \rightarrow Cr_2O_7^{2-} + H_2O$$

(v) Manganese atom has the structure $3d^54s^2$ and Mn^{2+} has $3d^5$, and hence manganous compounds are stable. The d^5 electrons can be removed by successive oxidation. This explains the oxidation stage of manganese in the range of 2+ to 7+.

In the 2+ oxidation state, the compounds are known as manganous compounds. In the 3+ oxidation state they are designated as manganic compounds. MnO_2 is an example of 4+ oxidation state and its salts are known as manganites like $CaMnO_3MnO_3$. Manganese trioxide exhibits 6+ oxidation state and its salts are known as manganates.

In the group oxidation state 7+ Mn_2O_7 is formed. Its salts are known as permanganates. The higher valencies are covalencies and not electrovalencies. This is quite in accordance with *Fajans* rule as covalence formation is encouraged by the small atomic volume of manganese and other transition element.]

(vi) The 4s electrons are removed in preference to 3d electrons as the latter are strongly stabilized as the nuclear charge increases. This stabilization also implies that more energy is required to remove a further electron to form the tripositive ion.

This explains the fact that after manganese which does exhibit the group oxidation state 7+, the dipositive oxidation is state except for ferric ions where the stability is attributed to $3d^5$ electronic configuration. With the increase in the effective nuclear charge the d electrons have been stabilized and in fact become core electrons. Iron exhibits valencies of 2, 3 and 6. In 2+ oxidation state ferrous compounds are formed. In 3+ oxidation state involving $3d^5$ electronic configuration, ferric compounds are obtained. In fact the stability of 3+ oxidation state is due to $3d^5$ configuration. In the 6+ oxidation state, the compounds formed are known as ferrates and they are derived from FeO_3.

(vii) In cobalt and nickel 2+ oxidation state is stable as already explained. Only $4s^2$ electrons are available here due to stabilization of 3d electrons due to increase in effective nuclear charge. However, they exhibit 3+ oxidation state also. The situation is interesting in the case of copper where $3d^{10}$ is complete. Hence to remove an electron from $3d^{10}$ for the formation of 2+ ion requires more energy.

HALIDES OF GALLIUM

Gallons Chloride, $GaCl_2$: It is prepared by heating gallic chloride with gallium.

$$2GaCl_3 + Ga \rightarrow 3GaCl_2$$

It is also prepared by heating gallium in chlorine or dissolving it in concentrated hydrochloric acid. It occurs as a white or pale grey, hygroscopic solid. In the fused state it conducts electric current. Strong solution of gallous chloride is powerful reducing agent. It is soluble in benzene.

Gallic Chloride, $GaCl_3$: Preparation : It is prepared by heating gallium at 240°C in an excess of chlorine or hydrochloric acid.

$$2Ga + 3Cl_2 \rightarrow 2GaCl_3$$

Properties

Anhydrous gallic chloride is very hygroscopic, fumes in the moist air. Gallic chloride when boiled with excess of water yields a white precipitate of oxychloride, GaOCl which redissolves on cooling.

Sulphates of Gallium

Gallous Sulphate, $GaSO_4$: It is formed when gallous oxide is dissolved in sulphuric acid. This sulphate does not form alums.

Gallic Sulphate, $Ga_2(SO_4)_3.H_2O$: It is formed by treating gallium, its oxide or hydroxide with sulphuric acid. On concentration, crystals of gallium sulphate are obtained.

It is soluble in water and its solution can be hydrolysed on boiling but the precipitate obtained redissolves on cooling. It is soluble in 60 percent alcohol but insoluble in ether. It forms alums with ammonium sulphate, potassium sulphate etc. Such as ammonium alum, $(NH_4)SO_4.Ga_2(SO_4)_3.24H_2O$ and potash alum, $K_2SO_4.Ga_2(SO_4)_3.24H_2O$.

Gallium nitrate, $Ga(NO_3)_2$: It is prepared by dissolving the metal, oxide or hydro-xide in nitric acid.

It occurs as a whit deliquescent substance. On ignition, it breaks up giving an oxide.

Analytical Reactions

Detection and estimation of gallium can be prepared either spectroscopically or by chemical methods.

1. Gallium salts yield precipitate with sodium or ammonium carbonate. In the excess of ammonium carbonate, precipitate is soluble.
2. Gallium salts with barium or calcium carbonate yield a precipitate of gallium hydroxide in cold.
3. If a boiling, dilute solution of a gallic salt, acidified with acetic acid, is heated with ammonium acetate, gallic hydroxide gets precipitated.

4. Gallium salts also give a precipitate with potassium ferrocyanide in strong hydrochloric acid.

Estimation

In gravimetric analysis, gallium can be precipitated by ammonia as hydroxide, which is ignited and weighed as the sesquioxide, Ga_2O_3.

CERIUM

Introduction : Cerium is one of the rare-earth elements and is included in that group of elements. Though it is in third group, its quadrivalency is well-established and in many ways it resembles thorium.

History

In 1803, the mineral cerite was analysed by ***Klaproth*** and ***Berzelius*** and it was found that mineral contained the new oxide ceria similar to yttria. Actually the oxide was found to contain later on a mixture of may rare-earth oxides. The metal cerium was isolated in 1826.

Occurrence

Cerium is most abundant of rare-earth metals. It occurs associated with other elements like iron, thorium etc.

1. **Euxenite :** It consists of niobates and tantalates of cerium earth.
2. **Bastnaesite :** It consists of fluocarbonate of cerium earth together with traces of thorium.
3. **Cerite :** It is hydrated silicate of cerium, $H_3(Ca,Fe)Ce_3Si_3O_{13}$, Ce: 51— 72%, Found in Sweden and Caucasus.
4. **Monazite :** It is a phosphate of cerium earth containing thorium and silica, Ce : 49—74%. Found in India, Brazil, Carolinas, Idaho and Scandinavia.
5. **Aeschnite :** It is a titano-niobate of cerium with thorium, Ce : 19.4—24.1%. Found in Norway, Germany, Brazil.
6. **Allanite :** It is hydrated silicate of cerium earth containing calcium, iron, beryllium, aluminium, Ce : 3.6 —51%. Found in Greenland and Scandinavia.

Extraction

For cerium monazite is the best source. Separation of cerium from other rare-earth elements, is based upon the fact that unlike other rare-earth elements cerium forms two distinct series of compounds derived from strongly basic $Ce(OH)_3$ and weakly, basic $Ce(OH)_4$. Ce(III) can be oxidised to Ce(IV) by means of $KBrO_3$., $KMnO_4$ etc which on boiling gives insoluble basic salt. The precipitation is fairly complete and adsorbs other rate-earths.

Monazite sand is an important source for the extraction of cerium. Monazite sand is treated with concentrated sulphuric acid. The reaction mixture is then lixivated with water and filtered. The rare-earths are converted into water soluble anhydrous rare-earth sulphate except thorium which is separated as insoluble residue. Part of thorium gone into the filtrate may be precipitated by the addition of sodium pyrophosphate and filtered out.

The filtrate is treated with oxalic acid to get the precipitates of rare-earths as oxalates. The precipitate is isolated by filtration and then boiled with alkali hydroxide the granular rare-earth hydroxides are formed which are washed and filtered. Cerium can be very readily separated from the other rare-earth by oxidation of Ce(III) to Ce(IV). Since ceric hydroxide is the weakest base of all and gets therefore, precipitated easily with dilute alkalies of weak bases.

The Ce(III) is oxidised in Ce(IV) with potassium permangnate, potassium persulphate, potassium bromate, chlorine, hydrogen peroxide or a mixture of nitric acid and ammonium nitrate as below :

1. Permanganate Method : In this method a neutral solution of nitrates is heated upto boiling and $KMnO_4$ is added till a pink colour appears. Cerium is then precipitated together with manganese, by adding the reagent consisting of $KMnO_4$ and Na_2CO_3 in the molar ratio.

$$3Ce(NO_3)_3 + KMnO_4 + 4Na_2CO_3 + 8H_2O \rightarrow 3Ce(OH) \downarrow + MnO_2.2H_2O + 8NaNO_3 + KNO_3 + 4CO_2 \downarrow$$

2. Bromate Method : The concentrated neutral solution of nitrates of rare earths is obtained by treating the oxide mixture of rare-earths with HNO_3. The solution of the nitrates is heated to boiling and potassium bromate is added. As soon as bromine fumes appear, lumps of marble are added and boiling is continued for 2-3 hours. When precipitate of basic cerium nitrate $[Ce(OH)(NO_3)_3]$ is obtained, boiling should be stopped. The marble is then removed and precipitate is allowed to settle. The precipitate adsorbs some other members of the rare-earth groups. The separated basic salt precipitate is again dissolved in HNO_3 and separation by bromate method is repeated once or twice.

3. Chlorine Method : In this method, Cl_2 is passed through the suspension of the hydroxides in an excess of caustic soda solution when $Ce(OH)_4$ remains undissolved while the other hydroxides dissolve as chlorides and hypo-chlorites. Several repetitions produce a satisfactory separation.

4. Ceric ammonium Nitrate Method : The basis of the method is that the double salt $Ce(NO_3)_4 2NH_4NO_3$ is much less soluble in HNO_3 than the corresponding salts of other rare earth-metals.

In this method the mixed oxides are dissolved in nitric acid and calculated quantity of ammonium nitrate is added to the hot solution. On cooling, the double salt-ceric ammonium nitrate separates as orange yellow crystals. It is purified by recrystallisation.

5. Electrolytic Method : It is the most efficient and rapid method of separating cerium from rare-earths. The solution of nitrates or sulphates is mixed with sodium phosphate and electrolysed when ceric phosphate is precipitated at the anode.

6. Solvent Extraction : A more recent method of separation consists in *solvent extraction* of ceric solutions in the tetrapositive in the tetrapositive valence state using tributyl phosphate (TBP) in mineral oil as the organic solvent.

The organic phase is stripped of its cerium content by reduction of Ce(IV) to Ce(III) by means of H_2O_2. The resultıng Ce(III) compounds, being insoluble in the organic phase, passes into the aqueous phase.

Cerium From Reactor Fuel

An additional source of cerium now-a-days is barium-140, recovered from used up reactor fuels. It is done by treating the latter with NaOH solution, which dissolves aluminium, leaving uranium and most of the fission products in the residue.

The insoluble portion is dissolved in HNO_3 and the solution concentrated to small bulk. Fuming HNO_3 is now added, when barium gets precipitated as barium nitrate which contains some strontium nitrate as contaminant.

The impure salt is dissolved and the solution treated with K_2CrO_4 at controlled pH, when $BaCrO_4$ gets precipitated, while strontium is left behind in solution. Repetition of the procedure gives a fairly pure product.

Barium- 140 obtained from the nitrate, is β-active and undergoes the following decay chain to produce cerium-140.

$$^{140}_{56}\text{Ba} \xrightarrow[12.8 \text{ days}]{-\beta} {}^{140}_{57}\text{La} \xrightarrow[40.2 \text{ hrs.}]{-\beta} {}^{140}_{58}\text{Ce (stable)}$$

Metallic Cerium

(a) Metallic cerium was first prepared by reduction of ceric chloride with sodium.

$$CeCl_4 + 4Na \rightarrow Ce + 4NaCl$$

But the yield is poor and is contaminated with NaCl.

Similarly, reduction of fluoride with calcium or aluminium yields an impure metal.

(b) The metal is commercially prepared by the electrolysis of a fused mixture of chlorides of cerium, calcium and sodium. The mixture is melted in an iron pot, which is made the cathode during the electrolysis. The anode is an iron or carbon rod. Cerium in the molten condition, collects at the bottom of the pot, it is trapped and cast into plates.

Properties

Physical

(a) Cerium is steel-grey, lustrous metal.

(b) It is malleable and ductile and is soft enough to cut with a knife.

(c) It is a poor conductor of electricity.

(d) It exists in four allotropic modifications.

(e) It produces sparks when struck with a piece of steel.

(f) It is paramagnetic.

(g) It melts at 795° C.

Chemical

Chemically, cerium is quite active.

1. Action of Air : Cerium is stable in dry air but it tarnishes in most air. It has low ignition temperature and catches fire at 160°C. On burning it produces ceric oxide (CeO_2) with much heat of formation.

2. Action of Water : Cold has little action on cerium but cerium evolves H_2 from boiling water.

$$2Ce + 6HCl \rightarrow 2Ce(OH)_3 + 3H_2$$

3. Action of acid : Dilute and concentrated HCl and HNO_3 attack the metal at room temperature.

$$2Ce + 6HCl \rightarrow 2CeCl_3 + 3H_2$$

$$3Ce + 3HNO_3 \rightarrow 3CeO_2 + 4NO + 2H_2O$$

Only dilute H_2SO_4 attacks the metal while concentrated H_2SO_4 has practically no action.

$$2Ce + 3H_2SO_4 \rightarrow Ce_2(SO_4) + 3H_2$$

4. Combination with Non-metals : Cerium directly combines with H_2 at 345°C and forms bluish-black hydride which is readily decomposed at high temperatures. The hydride is of interstitial type. It burns with Cl_2 at 220°C forming its chloride, $CeCl_3$. It also combines with I_2, S, As, Se, Te, P, C, N_2, etc. at different temperatures to form the binary compounds.

5. Reducing Action : Cerium reduces CO, CO_2 and CCl_4, with liberation of carbon. For this reason, CO_2 and CCl_4 are ineffective in putting out cerium fires.

Uses

1. Pure metallic cerium has no commercial uses but in the form of its alloys it has interesting applications. The important alloy of cerium is *misch metal.* It is used in gas and cigarette lighters.
2. Addition of a small proportion of cerium or misch metal to stainless steel improves their workability remarkably.
3. Addition of misch metal to steel has been found to increase the strength as much as 25 percent.
4. Magnesium alloys containing misch metal (3%) and zirconium (1%) having sufficiently high strength at elevated temperatures have been found particularly useful in making jet engine parts.
5. Cerium carbide and nitride are used in arc electrodes and in filaments for incandescent electric bulbs.
6. Cerium salts are used as a catalyst in the oxidation of hydrogen chloride to chlorine. Cerium hydride and nitride are used in the combination of N_2 and H_2 to form NH_3.

7. Ceric salts being oxidising agents are used in the manufacture of aldehydes, quinones, etc. from aromatic hydrocarbons.
8. Ceric salts are used as a mordant in dyeing cotton.
9. It is also used in glass for spectacles.

Detection

(i) *Small quantities of cerium can be detected by adding the neutral solution to a warm concentrated potassium carbonate solution and then adding a few drops of dilute hydrogen peroxide. A yellow colour is produced.*

(ii) *If the solution of a cerous salt when treated with ammonium hydroxide, an intense red-brown precipitate is formed.*

(iii) *A solution of a ceric salt when treated with ammonium hydroxide or sodium hydroxide yields yellowish precipitate.*

(iv) *A very sensitive test for cerium (III) is provided by an ammoniacal silver nitrate. The reagent is added to a neutral solution, when a mixed precipitate, consisting of Ce(IV) hydroxide (Yellow) and finely divided metallic silver (black) is formed.*

$$Ce^{3+} + [Ag(NH_3)_2]^+ + 4OH^- \rightarrow Ce(OH)_4 + Ag + 2NH_3$$

The test should be performed in the absence of Fe, CO *and* Ni.

(v) ***Dry test :*** *When a cerium is fused in to a borax bead in an oxidising flame, the bead is dark brown when hot and pale yellow to colourless when cold. In the reducing flame, the bead becomes colourless.*

Spot Tests

(a) Test For Ce^{3+} Compounds

(i) ***Phosphomolybdic acid test :*** *A drop of the test solution and of phosphomolybdic acid solution are mixed on a spot plate, and a drop of 40% sodium hydroxide is added. A intense blue colour or precipitate appears.*

(ii) ***Ammonaical silver nitrate test :*** *As described above in test number in detection (iv).*

(b) Test For Ce^{4+} Compounds

(i) ***Benzidine test :*** *A drop of the test solution is placed on filter paper, followed by a drop of dilute sodium hydroxide, and by a drop of benzidine solution. A deep blue colour develops (Blazejak-Ritges, 1970).*

(ii) **4′. *Amino acetophenone test :*** *This reagent reacts with* Ce^{4+} *compounds in 1.5 M HCl or* H_2SO_4 *solution giving a purple colour which changes to a red-brown precipitate (Somidevamma 1947).*

Estimation

(i) ***Gravimetric estimation :*** *In a ceric salt, cerium can be estimated gravimetrically by precipitation as ceric iodate,* $Ce(IO_3)_4$*, by using a solution of* KIO_3*. The precipitate is boiled with water and oxalic acid until all iodine has been expelled.*

It is then filtered, washed, dried, ignited and weighed as cerium dioxide, CeO_2*. In the case of cerous salts, cerous is converted into ceric by oxidation with potassium bromate and concentrated nitric acid, before precipitation as the iodate.*

(ii) ***Volumetric estimation :*** *Cerium in a ceric salt can be readily estimated by adding to a measured volume of the ceric solution acidified with sulphuric acid, N/20 ferrous ammonium sulphate solution in sufficient excess (as shown by complete discharge of yellow colour of ceric ion and back titrating the excess of ferrous ion present by means of standard* $KMnO_4$*).*

In the case of a cerous salt, the solution is first submitted to oxidation to the ceric state by means of sodium bismuthate in the presence of sulphuric acid. A measured volume of the oxidised solution is then subjected the above procedure.

Compounds

Cerium is known to form two classes of compounds. Cerous compounds, in which the oxidation number of cerium is +3, by the loss of there electron *i.e.,* two from 4f-orbitals ; hence the electronic configuration of Ce(III) ion would be $[Kr]4f^1$. Ceric compounds, in which the oxidation number of cerium is +4, by the loss of four electrons, *i.e.,* two from 6s and two from 4f-orbitals; hence the electronic configuration of Ce(IV) ion is [Kr].

CEROUS OXALATE $CE_2(C_2O_4)_3.10H_2O$

Preparation

It is obtained when oxalic acid is added to solution of cerous salt.

$$2Ce(NO_3)_3 + 3H_2C_2O_4 \rightarrow Ce_2(C_2O_4)_3 + 6HNO_3$$

Properties

It is a white crystalline powder, insoluble in water. It dissolves only to a slight extent in dilute strong acids, a behaviour that is typical of all the rare-earth oxalates.

Uses

It finds use medicinally as a sedative in sickness, particularly that of pregnancy, also in chronic-diarrhoea, hysteria, epilepsy and migraine.

Ceric Sulphate $Ce(SO4)_2$

Properties

It is obtained by heating a mixture of ceric oxide and concentrated sulphuric acid on a sand bath. The product is washed with ice-cold water and dried over caustic alkali.

$$CeO_2 + 2H_2SO_4 \rightarrow Ce(SO)_2 + 2H_2O$$

Properties

It is a deep orange yellow crystalline powder It decomposes on heating at above 195°C. It dissolves in water yielding a brown-yellow solution from which salt separates below 3°C as the do decahydrate, between 3° and 42.5° as the octahydrate and above 42.5°C as the tetrahydrate.

On diluting an aqueous, solution of ceric sulphate hydrolysis takes place and ultimately, sparingly soluble basic salts get precipitated.

It is a good oxidising agent for the oxidation of potassium nitrite, formic acid, acetic acid, tartaric acid, oxalic acid, methyl alcohol, hydrazine, hydroxylamine etc.

$$2Ce(SO_4)_2 + H_2C_2O_4 \rightarrow Ce_2(SO_4)_3 + H_2SO_4 + 2CO_2 \uparrow$$

$$2Ce(SO_4)_2 + 2N_2H_4 \rightarrow Ce_2(SO_4)_3 + (NH_4)_2SO_4 + N_2 \uparrow$$

$$2Ce(SO_4)_2 + Na_2S_2O_3 \rightarrow Ce_2(SO_4)_3 + Na_2SO_4 + Na_2S_4O_6$$

It forms double salts as follows :

$$2K_2SO_4.Ce(SO_4)_2.2H_2O ; 3(NH_4)_2SO_4. xCe(SO_4)_2. yH_2O.$$

Uses

It finds use it volumetric analysis as an oxidising agent *e.g.*, for the titration of Fe^{2+}, $C_2O_4^{2-}$, NO_2^-, Cu^+, Mo^{5+}. U^{4+} and Te^{4+} ions.

Ceric Oxide, CeO_2 : It is prepared by heating cerous or ceric hydroxide, nitrate, sulphate and oxalate.

$$Ce(C_2O_4)_2 \rightarrow CeO_2 + 4CO \uparrow + O2 \uparrow$$

It is a white powder having yellowish tint. It turns to lemon-yellow on heating. It is stable oxide but vapourizes at 900°C in vacuum and melts at the temperature of the electric furnace. It is insoluble in acids except concentrated sulphuric acid.

$$CeO_2 + 2H_2SO_4 \rightarrow Ce(SO)_2 + 2H_2O$$

It is a very weak basic oxide. It reacts with uranium dioxide yielding deep blue cubic crystals of $UO_2.CeO_2$.

Uses

It is used in highly efficient glass polishing compositions.

Ammonium Hexanitratocerate (IV), $(NH_4)[Ce(NO_3)_6]$

Preparation

This is commonly known as ceric ammonium nitrate. It is obtained by evaporating a mixed solution of ceric hydroxy nitrate, $Ce(OH)NO_3)_3$. 4½H_2O and ammonium nitrate in concentrated nitric acid.

$$Ce(OH)(NO_3)_3 + 2NH_4NO_3 \xrightarrow{HNO_3} (NH_4)_2[Ce(No_3)_6]$$

Properties

The Compound is a red crystalline solid. It is hydrolysed in water giving hydrosol and hydrogen. In acidic medium it is a strong oxidising agent. In the reduction of ceric salts in acid solutions with reducing agents, the single valency change occurs

$$Ce^{4+} + e \rightleftarrows Ce^{3+}$$

The equivalent weight of a ceric salt is therefore the molecular weight. In many volumetric redox titrations ammonium hexanitratocerate in the presence of H_2SO_4 can be employed for which ceric sulphate is used. Ammonium hexanitrato oxidises H_3AsO_3 to H_3AsO_4, Fe^{2+} to Fe^{3+}, $[Fe(CN)_6]^{4+}$ to $[Fe(CN)_6]^{3-}$.

$$2Ce^{4+} + H_3AsO_3 + H_2O \rightarrow 2Ce^{3+} + H_3AsO_4 + 2H^+$$

$$Ce^{4+} + Fe^{2+} \rightarrow Ce^{3+} + Fe3^+$$

$$Ce^{4+} + [Fe(CN)_6]^{4-} \rightarrow Ce^{3+} + [Fe(CN)_6]^{3-}$$

It also oxidises oxalic acid to CO_2.

$$H_2C_2O_4 + 2Ce^{4+} \rightarrow 2CO_2 + 2Ce^{3+} + 2H^+$$

Potassium nitrite is oxidised to KNO_3.

$$KNO_2 + 2Ce^{4+} + H_2O \rightarrow 2Ce^{3+} + KNO_3 + 2H^+$$

INDIUM

Discovery

In 1863 indium was discovered by? *Reich* and *Richter*, while investigating sphalerite ore sepctroscopically. They observed prominent indigo blue lines which were never noticed prior to this investigation. On the basis, of the indigo-blue line, the new element was named "indium".

Occurrence

Indium does not occur in the native state. In the combined state it occurs very widely in nature in small quantities. Its main source is Zinc blende in which it occurs to the extent of 0.1 percent. Indium almost occurs invariably in minute amount in cassiterite (tin stone), and also often found in pyrites. In traces, it also occurs in smithsonite and calamine. Lockyer reported the presence of indium lines in the solar spectrum.

Extraction

There are number of methods for the extraction of indium from various sources. Some of the important ones can be discussed as follows:

1. Reich and Richter's Method from Zinc Blende

Zinc blende is powdered and dissolved in nitric acid. Hydrogen sulphide gas is allowed to pass through the solution, when heavy metals like Cu, Pb, As, Cd and Mo are precipitated which are filtered out. The filtrate is treated with an excess of ammonia solution when most of the zinc goes into solution while indium and some other hydroxides are precipitated.

The precipitate is first of all washed and then dissolved in acetic acid and hydrogen sulphide is passed into the solution, which precipitates indium sulphide and also the sulphides of Fe, Zn and Mn etc. The ppt is now again dissolved in hydrochloric acid and reprecipitated with an excess of ammonium hydroxide. This is repeated so as to remove completely zinc and manganese. The final precipitate of indium hydroxide is dissolved in acetic acid and treated with hydrogen sulphide, when pure beautiful yellow precipitate of indium sulphide is obtained.

$$In(OH)_3 + 3CH_3COOH \rightarrow In(CH_3COO)_3 + 3H_2O$$

$$2In\ (CH_3COO)_3 + 3H_2S \rightarrow In_2S_3 + 6CH_3COOH$$

2. Winkler's Method

The indium having zinc is treated with dilute sulphuric acid insufficient to dissolve all the metal. On the residue of undissolved zinc, a spongy mass of lead, copper, tin, arsenic, iron and indium gets deposited. The spongy material is collected and mixed with strong sulphuric acid and heated strongly so as to volatilize most of the acid. To this solution an excess of ammonia is added which precipitates indium hydroxide along with the hydroxides of Fe, Zn and cadmium.

This precipitate is washed and made to dissolve in a little hydrochloric acid and the ferric iron is reduced to ferrous by sulphur dioxide. The solution is now treated with barium carbonate when indium gets precipitated as hydroxide.

Metallurgy

1. Indium can be obtained by reducing its oxide with hydrogen, carbon or coal gas (Reich and Richter),

$$In_2O_3 + 3H_2 \rightarrow 2In + 3H_2O$$

$$In_2O_3 + 3C \rightarrow 2In + 3CO$$

2. It is also obtained by Winkler's method by heating the oxide or chloride of indium with metallic sodium under a layer of dry sodium chloride.

$$In_2O_3 + 6Na \rightarrow 2In + 3Na_2O$$

$$InCl_3 + 3Na \rightarrow 2In + 3Na$$

3. Rossler and Wolf obtained indium by the addition of zinc shavings to a slightly acid solution of an indium salt ($InCl_3$), when metallic indium is precipitated.

$$2InCl_3 + 3Zn \rightarrow 2In + 3ZnCl_2$$

4. Schucht obtained indium by the electrolysis of indium sulphate or chloride or nitrate in the presence of pyridine, hydroxylamine or formic acid.

Properties

Physical

Indium occurs as a silvery white metal like that of platinum. The electro-deposited metal forms octahedral crystals belonging to cubic system, the crystals being isomorphous with gallium and aluminum. As indium is a very soft metal, it readily marks paper. It is a ductile metal. It melts at 155°C. Indium salts impart bluish-red colour to the flame.

Chemical Action of Air

At ordinary temperature, indium is quite stable in air, and retains it lusture. When it is heated above its melting point, a film of monoxide, InO is formed and on heating further at higher temperature it yields indium trioxide, In_2O_3.

Action of halogens : It directly reacts with halogens halides.

Action with sulphur : With sulphur on heating, the combination occurs with incandescence. Action of acids : It dissolves slowly in cold dilute mineral acids, but more rapidly when heated, with the evolution of hydrogen.

$$2In + 6HCl \rightarrow 2InCl_3 + 3H_2$$

Indium dissolves in a solution of oxalic acid but not in acetic acid

Action of alkalies : Indium is not effected on treating with potassium hydroxide.

Action with other metals : indium readily alloys with sodium gold, tin, lead gallium and platinum.

Uses

Indium does not find much use. It is having the characteristic to harden to provide strength to the metal with which it is alloyed. The alloys of indium are highly resistant to corrosion and do not tarnish easily and thus find use in jewellery and dental work. Indium compounds are used for colouring glass, paints, fluorescent lighting etc.

Indium compounds

Oxides of Indium. Indium is known to form number of oxides. Its hemioxide, In_2O is a black coloured solid. It is obtained by reducing the trioxide, *i.e.,* by heating the trioxide in hydrogen at 500°C. followed by sublimation in a high vacuum at about 700°C. It acts as a powerful reducing agent and liberates hydrogen from acids. On heating in air it is readily oxidized into sesquioxde.

Indium monoxide, In_2O_3 is the stable oxide. It is formed if the metal burns in oxygen, or if the hydroxide or a salt with a volatile acid is ignited. Indium sesquioxide is yellow at room temperature, but brown when hot.

Halides of indium : Indium is known to form number of halides such as fluorides, chlorides, bromides and iodides, but its most important halide indium trichloride $InCl_3$ is described here.

Indium Trichloride, $InCl_3$: Preparation

1. It is prepared by heating indium dichloride in a current of chlorine.
2. It is also obtained by heating the trioxide and carbon mixture in a current of chlorine.

   $In_2O_3 + 3C + 3Cl_2 \rightarrow 2InCl_3 + 3CO$
3. Another method of its preparation is by heating the metal in moist hydrochloric acid gas at 120-130°C.

Properties

It exists as colourless crystalline substance. It melts at 586°C into a pale-yellow liquid. Its Sp. Gr. is 4.0. It is a deliquescent substance and dissolves in water. If its aqueous solution is evaporated at a high temperature, a white powder of indium oxychloride, InOCl is obtained. Indium chloride acts as a catalytic agent in chlorinating benzene.

Indium sulphate, $In_2(SO_4)_3.9H_2O$: It is obtained by dissolving the oxide, hydroxide, or carbonate of indium in sulphuric acid, and evaporating the solution to almost to a syrup. This is then cooled in a desiccator when deliquescent crystals of hydrosulphate, are obtained.

When dried at 100°C, the sulphate $In_2(SO_4)_3.9H_2O$ is obtained as a white powder. It exists as a white powder. It also forms double sulphates and alums with ammonium sulphate, rubidium sulphate etc.

$$(NH_4)2SO_4.In_2(SO_4)_3.In_2(SO_4)3.24H_2O; Rb_2So_4.In_2(SO_4)_2.24H_2O$$

Analytical Reaction

1. Indium salts with sodium or ammonium carbonate yield a gelatinous precipitate of indium carbonate. This precipitate of indium carbonate is soluble in excess of ammonium of ammonium carbonate but reprecipitates on boiling.
2. With oxalic acid and alkali oxalates, indium salts yield crystalline precipitation of indium oxalate.
3. Indium salts yield a whit precipitate with potassium ferrocyanide and a yellow precipitate with potassium chromates.
4. Quantitatively, indium salts can be precipitated with organic bases like dimethyl aniline, piperidine, quinoline etc.
5. Indium salts yields white precipitate of indium hydroxide with a solution of hydroxylamine.
6. Indium can also be detected by using either alizarin or quinalizarin. One or two drops of neutral or slightly acidic indium salt solution is kept on a filter paper which is previously impregnated with a saturated alcoholic solution of alizarin or quinalizarin. The moist paper is kept over ammonia and then finally immersed in a saturated queous solution of boric acid. Red or violet against the yellow background of paper, reveals the presence of indium.

Estimation

Quantitatively it could be estimated by precipitating with ammonia as hydroxide and would be weighed as oxide after subsequent ignition

COMPLEX FORMATION

The ability to form complexes is in marked contrast to the s- and p-block elements which from only a few complexes. The reason transition elements are so good at forming complexes is that they have small, highly charged ions and have vacant low energy orbitals to accept lone pairs of electrons donated by other groups or ligands. Complexes where the metal is in the (+III) oxidation state are generally more stable than those where the metal is in the (+III) state.

The central metal atom or ion which exercises two different valencies, *i.e.,* electrovalency and dative covalency or around which the ligands are arranged is usually the atom or ion of the transition element. Of course, all metal ions including those of the most electropositive elements form complexes. But it must be noted that the ions of the *more weakly electropositive metals weakly electropositive metals (transition elements) form more stable complexes with a wider range of ligands than those of the s-block elements.* Thus the weakly electropositive character of the transition elements is associated with the stability of the complex compounds. This is partly due to the high effective nuclear charge which transition metal ions exert at bonding distances from the nuclei, increasing their attraction for ligands.

The transition metal ions are relatively small. Besides, due to the presence of poorly shielding d electrons, the nuclear charge attracting the ligands is higher than expected and they attract ligands more than non-transition metal ions of the same size and overall charge and hence form more stable complexes. In most cases, though not as a rule the effective atomic number *i.e.,* the number of electrons in the central metal ion and the electrons donated by the ligands together is normally same as the atomic number of the neighbouring inert gas, *e.g.,* $K_4[Fe(CN)_6]$, $Ni(Co)_4$. etc.

The *weak ligands like* Co, No, C_2H_4 etc. *form complexes only when transition metals are in zero of low oxidation states.* This is attributed to the fact that the donor atom of thee ligands is having vacant orbitals in addition to the lone pair. Thus such ligands form following types of bondings.

(i) They first of all form a normal s-Covalent by donating its electron pair to the metal atom.

(ii) Then the vacant orbitals of the ligand accept electrons from the filled metal orbitals to form a p-bond (dp-pp overlap) which addes to the strength of the s-bond. Thus it refers to the ability of the ligand to form s-bond (back acceptance of metal electrons) that is able to stabilise the s-bond between metal and ligand and hence the complex molecule. The p-bond formation in turn becomes possible when the metal atom is having atom is having filled d-orbitals *i.e.,* when it is in a low or zero oxidation state.

On the other hand, *the highly electronegative and basic ligands* such as F, Cl, NH_3, etc. *form complexes with transition metals even when they are in high oxidation states*. The reason is that the small, highly charged or neutral ligands having a lone pair of electrons may form strong σ-bond by donating a lone pair of electrons. The complexes formed are stable enough to be having an independent existence. This is attributed to greater attraction of metal ion for ligand electrons. This explains why Co^{3+} ion forms a stable complex with ammonia, $[Co(NH_3)_6]^{3+}$ while Co^{2+} does not form stable complex with ammonia.

Trends in Transition Series for Complex Formation

(i) In each transition series, the stability of complexes tends to increase with increasing atomic number of the element and in a particular oxidation state with decreasing size of its atoms.

(ii) If the transition metal atom exhibits more than one oxidation state, the higher valent ion is able to form more stable complexes.

Colour

Many ionic and covalent compounds of transition elements are coloured. In contrast compounds of the s-and p-block elements are almost always white. When light passes through a material. It is deprived of those wavelengths that are absorbed. If absorption occurs in the visible region of the spectrum, the transmitted light is coloured with the complementary colour to the colour of the light absorbed. Absorption in the visible and UV regions of the spectrum is caused by changes in electronic energy. Thus the spectra are sometimes called electronic spectra. (These changes are often accompanied by much smaller changes in vibrational and rotational energy. It is always possible to promote an electron from one energy level to another. However, the energy jumps are usually so large that the absorption lies in the UV region. Special circumstances can make it possible to obtain small jumps in electronic energy which appear at absorption in the visible region.

Polarization

NaCl, NaBr and NaI are all ionic, and are all colourless. AgCl is also colourless. Thus the halide ions Cl^-, Br^- and I^-, and the metal ions Na^+ and Ag^+, are typically colourless. However, AgBr is pale yellow and AgI is yellow. The colour arises because the Ag^+ ion polarizes the halide ions. This means that it distorts the electron cloud, and implies a greater covalent contribution. The polarizability of ions increases with size : thus I^- is the most polarized and is the most coloured. For the same reason Ag_2CO_3 and Ag_3PO_4 are yellow, and Ag_2O and Ag_2S are black.

Incompletely Filled d or f Subshell

Colour may arise from an entirely different cause in ions with incomplete d or f shells. This source of colour is very important in most of the transition metal ions. The colours and the number of 3d-electrons present in some hydrated ions of the transition metals of the first transition series are given in Table 1.8.

It is significant to note from Table 1.8 that hydrated Sc^{3+} ions having $3d^o$ and hydrated Zn^{2+} ions having $3d^{10}$ electrons have been colourless. This clearly reveals that the colour is associated with $3d^1$ to $3d^9$ electrons, *i.e.,* with the transition ions. In fact the explanation for the colour of a transition metal ion has been a complex one but the same could be explained below in a simple manner with suitable examples.

Table 7 : Colours of Hydrated ions of the first Transition series

Ion (aq)	*Outer electronic configuration*	*No. of unpaired electrons*	*Colour*	*Magnetic moment (BM)*
Sc^{3+}	$3d^0$	0	colourless	0
Ti^{3+}	$3d^1$	1	purple	1.75
V^{3+}	$3d^2$	2	green	3.86
C_1^{3+}	$3d^3$	3	violet	4.80
Mn^{3+}	$3d^4$	4	violet	5.92
Fe^{3+}	$3d^5$	5	yellow	5.96
Mn^{2+}	$3d^5$	5	pink	5.0–5.5
Fe_{2+}	$3d^6$	4	green	4.4–5.2
Co^{2+}	$3d^7$	3	pink	2.9
Ni^{2+}	$3d_8$	2	green	2.9–3.4
Cu^{2+}	$3d^9$	1	blue	1.4–2.2
Cu^+	$3d^{10}$	0	colourless	0
Zn^{2+}	$3d^{10}$	0	colourless	0

The colour of a transition metal ion is related to the following factors:

(i) Incomplete d orbitals (between 1 and 9 electrons).

(ii) The nature of ligands attached to the central metal ion.

Colour could be developed due to the movement of electrons from one d orbital to another.

We will now describe some examples of development of colour on the basis of promotion of electrons from one d level to another d level with fairly small energy difference. When light passes through a solution of $[Ti(H_2O)_6]^{3+}$ ions, a quantum of light may be absorbed by the ion and this excites an electron from a lower-level d orbital. The frequency v of the light absorbed is given by the equation $\Delta E = hv$, where h is the Planck's constant and ΔE the energy difference due to d–d transition of electrons.

The energy difference for $[Ti(H_2O)_6]^{3+}$ has been such that the light absorbed is in the visible region of the spectrum and the ion is coloured. In fact green light (wave length about 5000 Å) is absorbed *i.e.,* white light minus green light gives purple light. Similarly, in the case of $[Cu(H_2O)_6]^{2+}$ ion the energy difference is illustrated in Fig. 1.1a. In fact two of the 3d orbitals are raised in energy as compared to the other 3d orbitals. The degeneracy (equal energy) of all the 3d orbital is thus destroyed due to the repulsion of 3d electrons of the cupric ion which are close to those of the ligands, *i.e.,* water molecules in $[Cu(H_2O_6]^{2+}$. The energy difference is given by the following Planck's equation.

$$\Delta E = hv.$$

The relation of energy of the frequency given by the above equation excites an electron which moves from lower energy level to higher energy level. In the case of hydrated Cu^{2+} ions, red light of the appropriate frequency from the *visible spectrum* is absorbed due to the transition of electrons in d–d levels involving small energy differences. Since white light minus red gives blue, the colour of hydrated Cu^{2+} ions is blue. In fact, it is interesting to observe that the pale blue hydrated Cu2+ ion changes to deep blue colour in the presence of excess of ammonia and to green colour in the presence of excess of chloride ions.

$$\underset{\text{pale blue}}{[Cu(H_2O)_6]^{2+}} + 4Cl^- \rightarrow \underset{\text{green}}{[CuCl_4]^{2-}} + 6H_2O$$

It is generally found that the ions of the elements of the first transition series absorb light from the *visible* region of the spectrum, hence variety of colours are observed. The exact colour, depends on the frequency of light absorbed.

Some compounds of the transition metals are white, for example $ZnSO_4$ and TiO_2. In these compounds it is possible to promote electrons within the d level. Zn^{2+} has d^{10} configuration and the d level is full. Ti^{4+} has d^{o} configuration and the d level is empty. In the series Sc(+III), Ti (+IV), V(+V), Cr(+ VI) and Mn(+VII), these ions may all be considered to have an empty d shell.

Hence, d–d spectra are impossible and they should be colourless. However, as the oxidation number increases these states become increasingly covalent. Rather than form highly charged simple ions, oxoions are formed TiO^{2+}, VO^{+2}, VO_4^{3-}, CrO_4^{2-} and MnO_4^-. VO_4^-, VO_2^+ is pale yellow, but CrO_4^{2-} is strongly yellow coloured, and MnO_4^- has an intense purple colour in solution though the solid is almost black. The colour arises by charge transfer.

In MnO_4^- an electron is momentarily transferred from O to the metal, thus momentarily changing O^{2-} to O^- and reducing the oxidation state of the metal from Mn(VII) to Mn(VI). Charge transfer requires that the energy levels on the two different atoms are fairly close. Charge transfer always produces intense colours since the restrictions of the Laporte and spin selection rules do not apply to transitions between atoms.

In contrast, the ions of s and p block elements are usually colourless. In the case of these ions, there can not be d–d transition and the energy required to promote s or p electrons to a higher energy level is much greater and may correspond to ultraviolet region, hence their compound appear colourless. So ions of these elements are either yellow or colourless.

Reactivity

Due to the small size of their atoms, ionization energies of d-block have been fairly high. When these elements form metal ion, the latter do not get hydrated to the presence of covalent bonding. So a larger amount of energy is needed for change from solid to vapour state.

The d-block elements have been unreactive due to following factors.

(i) **High ionisation energies :** Because of the small size of their atoms, ionisation energies of the d-block elements are fairly high.

(ii) **Low heats of hydration :** The metal ions do not get hydrated easily.

(iii) **High heats of sublimation :** *Due to the presence of covalent bonding, these are having high heats of sublimation.*

The tendency to remain unreactive has been found to be more pronounced in platinum and gold in the third transition series.

Standard Electrode Potential

Except copper and mercury, the standard reduction potential of d-block elements in acid solution is generally negative, *i.e.,* lower than that of hydrogen (taken as zero). However copper and mercury are having positive electrode potentials. All the d-block elements with negative reduction liberate hydrogen from dilute acid.

$$M + 2H^+ \rightarrow M^{2+} + H_2(g)$$

The rates of evolution of hydrogen have been found to be very low in some of the case. Some of these metals have been unreactive because of the formation of the protective film of oxide on their surface. For example, chromium is so unreactive that it can be used as a protective non-oxidising metal.

Reducing Properties

Transition metal with sufficiently negative standard reduction potentials must act as good reducing agents *i.e.,* they should be oxidised easily to their ions. However, they have been not good reducing agents when compared with elements of groups I, II and III.

This is so as the reducing property of a metal depends on its tendency to form ions. The transition metals have less tendency to form ions because of (i) their high ionization energies (ii) low heat of hydration of their ions and (iii) high heats of sublimation.

$$M \rightarrow M^{2+} + 2e^-$$

Magnetic Behaviour

As electric current flows through a wire, magnetic moment is generated. Similarly the electrons spin on their axes and are regarded to generate magnetic moment. The electrons occupying the same orbital have zero magnetic moment as the opposite spins of the two electrons counter the magnetic moment.

Substances which are weakly repelled by the strong magnetic field are termed as *diamagnetic* while those which are weakly attracted by a strong magnetic field are termed as *paramagnetic.*

On the basis of magnetic properties, substances are classified into the following two types.

(i) *Paramagnetic substances :* Substances which are weakly attracted by the magnetic field are termed as paramagnetic. These substances lose their magnetism on removing the magnetic field. Paramagnetism is produced by the presence of unpaired electrons and as most of the transition metal atoms are having unpaired d-electrons, they are paramagnetic in behaviour.

(ii) *Dimagnetic substances :* Substances which are repelled by magnetic field are termed as diamagnetic. It is the property of the completely filled electronic subshells. As all elements (except hydrogen atom) are having filled electronic shells, some diamagnetism is shown by all substances. A simple example of a diamagnetic substance is sodium chloride.

The magnetic moment increases with the number of unpaired electrons. In 3d series, the magnetic moment, μ 3d (in Bohr magneton, BM) is related to the number of unpaired electrons, n by the expression

$$\mu = \sqrt{[n(n+2)]}$$

Based on this expression, the magnetic moments can be calculated and compared with those obtained experimentally (Table 1.10). It is clear that the metal ion with the largest number of unpaired electrons will have the maximum magnetic moment.

Mn^{2+} ($3d^5$) has 5 unpaired electrons. Its calculated magnetic moment will be,

$$\mu = \sqrt{[n(n+2)]}$$
$$= \sqrt{[5(5+2)} = \sqrt{[5(7)]} = 5.92 \text{ B.M.}$$

$$\text{*(BM)} = \frac{eh}{4\pi mc}$$

where e is the electronic charge,

h is the Planck's constant.

m is the mass of electron.

and c is the speed of light.

Fe^{2+} ($3d^6$) has 4 unpaired electrons. Its magnetic moments will be

$$\mu = \sqrt{[n(n+2)]}$$

$$= \sqrt{4(4+2)]} = \sqrt{4(6)} = 4.90 \text{ B. M.}$$

Thus, Mn^{2+} will have a higher magnetic moment.

Table 8 : Calculated and Observed Magnetic Moments of some ions of 3d Series

Ion	*Outer electronic configuration*	*No. of unpaired electrons*	*Magnetic moments (BM)* calculated*	*Observed*
Sc^{3+}	$3d^o$	0	0	Diamagnetic
Ti^{3+}	$3d^1$	1	1.73	1.75
Ti^{2+}	$3d^2$	2	2.84	2.76
V^{2+}	$3d^3$	3	3.87	3.86
Cr^{2+}	$3d^4$	4	4.90	4.80
Mn^{2+}	$3d^5$	5	5.92	5.96
Fe^{2+}	$3d^6$	4	4.90	5.0–5.5
Co^{2+}	$3d^7$	3	3.87	4.4–5.2
Ni^{2+}	$3d^8$	2	2.84	2.9–3.4
Cu^{2+}	$3d^9$	1	1.73	1.4–2.2
Zn^{2+}	$3d^{10}$	0	0	Diamagneyic

Compounds that display paramagnetism when kept in a magnetic field get pulled into the field; the extent of the pull can be measured by using the Gouy magnetic balance.

It is important to observe that diamagnetic substances exhibit a decrease in weight in the presence of a magnetic field.

Catalytic Behaviour

A number of transition metals and their compounds are known to catalyse various reactions of industrial importance *e.g.,* vanadium (V) oxide in the manufacture of sulphuric acid by Contact process, manganese (IV) oxide in the laboratory preparation of oxygen, finely divided iron in the Haber's process for the manufacture of ammonia, nickel in the hydrogenation of oils and copper (II) chloride in the manufacture of chlorine.

Many views have been advanced to explain the mechanism of catalysis. In the case of catalysis at a solid surface, bonds would be formed between the molecules of the reactants and the surface atoms of the catalyst. The 3d electrons of the first transition series elements are used in bonding in addition to 4s electrons. Thus, the concentration of the reactants at the catalyst surface increases and the bonds of the reactant molecules weaken because the activation energy is lowered.

The catalytic property of transition metal ions could be explained by their varying oxidation states, *eg.,* vanadium (V) oxide is first converted into vanadium (IV) oxide during the reaction but vanadium (V) oxide is recovered back at the end of the process as follows :

$$SO_2 + V_2O_5 \rightarrow SO_3 + V_2O_4$$

$$2SO_2 + V_2O_4 + O_2 \rightarrow 2VOSO_4 \quad \textit{Vandyl sulphate}$$

$$2VOSO_4 \rightarrow V_2O_5 + SO_3 + SO_2$$

$$3SO_2 + V_2O_5 + O_2 \rightarrow 2SO_3 + SO_2 + .V_2O_5$$

i.e., $$2SO_2 + O_2 \xrightarrow{V_2O_5} 2SO_3$$

Similarly, Fe^{3+} ions are able to catalyse the oxidation of iodides to iodine as follows :

$$2I^- + S_2O_8{}^{2-} \xrightarrow{Fe^{3+}} I_2 + 2SO_4{}^{2-}$$

One of the possible explanations for the above catalytic reaction may be

$$2Fe^{3+} + 2I^- \rightarrow 2Fe^{2+} + I_2$$

$$2Fe^{2+} S_2O_8{}^{2-} \rightarrow 2Fe^{3+} + 2SO_4{}^{2-}$$

Both the above reactions have been known to take place and it has been expected that two reactions between the ions of opposite charge would be faster than the one reaction between ions of the same charge.

Metallic Character

As the number of electrons in the outermost shell is very small, all the d-block elements have been metals. Unlike s-block metals, these have been hard, malleable and ductile. The melting and boiling points have been high. Their hardness along with their high melting and boiling points indicate the presence of strong binding forces in their crystals. The metals have lose packed structures (*fcc, hcp* or *bcc* type).

The d-block elements are good conductors of heat and electricity. However they differ from non-transition metals in being hard and brittle (note down the exceptional behaviour of mercury which is liquid and quite soft like alkali metals).

Metals of groups VIII and IB have been softer and more ductile. This could be explained in the light of their crystal structure. These crystallize to form *fcc* type crystal lattice while others crystallize with *hcp* of *bcc* structure. In the *fcc* type crystal lattice of these elements, their being larger number of planes where deformation can occur, these elements have been softer and more ductile.

Hardness of these metals indicates the presence of covalent bonds. The unpaired d-electrons may overlap to form covalent bonds. The number of such covalent bonds would be regarded as a measure of overall bond strength and hardness of the metal. In a transition series the number of unpaired d-electrons increase on moving from group IIIB [(n–1) d^1, ns^2] to VIB (n–1) [(n–1)d^5, ns^1] after which pairing begins.

Because of this the metallic lattice becomes stronger and stronger up to group VIB (Cr, Mo and W) and strength decreases progressively.

All these are good conductors of electricity and have metallic lustre. Both these properties could be explained on the basis of delocalization of their outer s-electrons over the entire crystal lattice.

Hard and brittle nature of these metals indicate *covalent bonding* which is also favoured by the presence of unfilled d-orbitals. On the other hand, the existence of *metallic bonding* is revealed by their good conductivity to heat and electricity. Thus it appears that covalent as well as metallic bonding exist amongst the transition elements.

MELTING POINTS AND BOILING POINTS

The d-block elements have high melting and boiling points. Transition elements typically melt above 1000°C. Ten elements melt above 2000°C and three melt above 3000°C (Ta 3000°C, W 3410°C and Re 3180°C). There are a few exceptions. The melting points of La and Ag are just under 1000°C (920°C and 961°C respectively). Other notable exceptions are Zn (420°C), Cd (321°C) and Hg which is liquid at room temperature and melts at –38°C. The last three behave atypically because the d shell is complete, and d electrons do not participate in metallic bonding. The high melting points are in marked contrast to the low melting points for the s-block metals Li (181°C) and Cs (29°C).

The high melting and boiling points of transition elements show that they are held by strong forces. This is due to the presence of (a) strong metallic bonding because of the overlapping of (n–1) d orbitals, and (b) covalent bonding by the unpaired d-orbital electrons. As Zn, Cd and Hg are having completely filled (n–1) d-orbitals, their atoms are not expected to form covalent bonding amongst themselves; hence they are having relatively low m.p. than the other d-block elements. In this region the elements have maximum number of unpaired d electrons available for bonding *i.e.,* metallic bond. The elements also exhibit the highest oxidation states in this region.

Atomic (Covalent) Radii

The variation in atomic radii across each transition series is not as simple as that observed in s- and p-block elements. Some overall trends in the variation of atomic radii, across the period have been as follows:

(i) The covalent radii of the (Table 1.1) decrease from left to right across a row in the transition series, until near the end when the size increases slightly. On passing from left to right, extra protons are placed in the nucleus and extra orbital electrons are added. The orbital electrons shield the nuclear charge incompletely (d electrons shield less efficiently than p electrons, which in turn are shielded less efficiently than pelectrons, which in turn are shielded less effectively than s electrons). Thus, the nuclear charge attracts all of the electrons more strongly hence a contraction in size occurs.

(ii) There occurs a small decrease in atomic radii when we pass from chromium to copper. It could be explained in terms of screening effect. Because successive addition of d-electrons screens the outer 4s electron from the inward nuclear pull. Due to this screening, the effective nuclear charge increases only

slightly and the size of the electron cloud does not therefore, change much on moving from chromium to copper.

(iii) In a particular transition series, the atomic radius attains a minimum value for the group VIII elements and increases again towards the end of the series. It could be explained in terms of the increased force of repulsion among the added electrons which becomes greater than the attractive force due to the increased nuclear charge and causes the expansion of the electron cloud.

Table 9 : Covalent radii of Transition Element (Å).

K	*Ca*	*Sc*	*Ti*	*V*	*Cr*	*Mn*	*Fe*	*Co*	*Ni*	*Cu*	*Zn*
1.57	1.74	1.44	1.32	1.22	1.17	1.17	1.17	1.16	1.15	1.17	1.25
Rb	Sr	Y	Zr	N	Mo	Tc	Ru	Rh	Pd	Ag	Cd
2.16	1.91	1.62	1.45	.34	1.29	–	1.24	1.25	1.28	1.34	1.41
Cs	Ba	La*	Hf	Ta	W	Re	Os	Ir	Pt	Au	Hg
2.35	1.98	1.69	1.44	1.34	1.30	1.28	1.26	1.26	1.29	1.34	1.41

–14 Lanthanide elements

The very close similarity between the radii of the elements of second and third transition series (*e.g.*, Zr and Hf; Nb and Ta; Mo and W; Tc and Re; etc.) has been a consequence of the filling of the 4f subshell.

The effects of the lanthanide contraction are less pronounced towards the right of the d-block. However, the effect still shows to a lesser degree in the p-block elements which follow.

Ionic Radii

The ionic radii of the d-block elements have been known to follow the same trend as their atomic radii as described above. For ions carrying identical charges, the ionic radius decreases slowly with increasing atomic number in the same transition series.

The radii of the doubly charged ions, although small, are comparable with the radius of Ca^{2+} ion (0.99Å). Thus the *oxides of these elements are similar to calcium oxide*, although somewhat less basic and less soluble in water.

Table 10 : Atomic and Ionic Radii (Å) of First Transition Series Elements

Radii	*Sc*	*Ti*	*V*	*Cr*	*Mn*	*Fe*	*Co*	*Ni*	*Cu*	*Zn*
Atomic	1.62	1.47	1.34	1.27	1.26	1.26	1.25	1.24	1.28	1.33
M2+	–	0.90	0.88	0.84	0.80	0.76	0.74	0.72	0.72	0.74
M3+	0.81	0.76	0.74	0.69	0.66	0.64	0.63	–	–	–

Atomic Volumes

Atomic volumes of the d-block elements have been much lower than those of the neighbouring s and p-block elements. Atomic volume in the same transition series first gets decreased and reaches a minimum with VIII group elements. It then starts increasing. This can be explained on the basis of the fact that in d-block elements the inner orbitals, *i.e,* (n–1) d, rather than the outermost np, become filled and thus the increased nuclear charge will be exerting more effect and pulls the electrons inward. This ultimately leads to decrease in atomic volume.

It is possible to explain this behaviour as in the case of atomic radii which follow the same trend.

Transition Series	*Ele-ment*	*At radius*	*At wt.*	*Density g cm*	*Ele-ment*	*At radius*	*At wt.*	*Density g cm*
Second	Nb	1.46	93	8.4	Mo	1.39	96	10.4
Third	Ta	1.46	181	16.6	W	1.39	184	19.3
Second	Tc	1.36	99	11.5	Ag	1.43	108	10.5
Third	Re	1.37	186	20.5	Au	1.44	197	19.3

Densities

The atomic volumes of the transition elements are low compared with elements in neighbouring groups I and II. This is because the increased nuclear charge is poorly screened and so attracts all the electrons more strongly. In addition, the extra electrons added occupy inner orbitals. Consequently the densities of the transition metals are high. Practically all have a density greater then 5g cm^{-3}. (The only exceptions are Se 3.0g cm^{-3} and Y and Ti 4.5g cm^{-3})

The densities of the second row are high and third row values are even higher. Two elements with the highest densities are osmium 22.57 g cm^{-3} and iridium 22.61 g cm^{-3}. To get some feel for how high this figure really is, a football made of osmium or iridium measuring 30 cm in diameter would weight 320 kg or almost one third of a tonne.

In a given transition series, when we move from left to right, the density first increases and then attains a maximum value on reaching group VIIII. Then it starts decreasing when we move further to group IB and IIB.

It is possible to explain *this in terms of their small radii and close packed structure.* Overall trend could be governed by the trend in the variation of atomic radii discussed earlier.

Density increases substantially when we move down a group. The atomic size of the elements in the same group but belonging to second and third transition series remains almost the same but their atomic weight is almost doubled.

Because of this the densities of elements in the third transition series have been almost double of the corresponding elements of second transition series, as seen from their values given below :

Chart Showing Variation of Densities of d-block Elements

IIIB IV VB VIB VIIB VIII IB IB

Increases horizontally Maximum value $\xrightarrow{\text{Decreases}}$ onwards

up to VIII group

Increases on moving down the group ↓ Increases

↓

Os

(maximum density 22.57 g.cm^{-3})

Ionization Potentials

As atoms of transition elements are small in size, their ionization energies are fairly high. The ionisation energies of most of the d-block elements lie in between those of s-and p-block elements. This reveals that the d-block elements are less electropositive than the s-block elements (alkali and alkaline earth metals).

Hence, d-block elements fail to form ionic compounds so readily as the alkali and alkaline earth metals. Further, unlike the s-block elements, the d-block elements are also having a tendency to form covalent compounds under certain conditions. Generally, lower valent compounds are ionic whereas the higher valent compounds are covalent.

Ionization energies increase on moving from left to right across a transition series. The increase, however, has been not so pronounced as in the case of s and p-block elements in the same period. In the case of first transition series the ionization energies of Sc, Ti, V and Cr have been almost close to one another. Similarly values for Fe, Co Ni and Cu have been almost similar.

It is possible to explain the above trend by considering the screening effect of the d-electrons. This is due to the fact that in case of transition elements the new electron is being added in (n–1) d subshell which provides extra screening effect and thereby shields the outer s-electrons from the inward nuclear pull.

In other words, the *effect of increasing nuclear charge is almost compensated by the extra screening effect of the nucleus provided by the increasing number of d-electrons. The* exceptionally higher ionisation energy value of zinc is attributed to its additional stability which is associated with completely filled 3d-orbital ($3d^{10}.4s^2$).

The values of their ionization energies have been somewhat between those of s-block elements on their left and p-block elements on their right. It means that these have been less electropositive than the s-block elements. They are able to from ionic compounds but not so readily as the s-block elements. Unlike s-block elements d-block elements are able to form covalent compounds.

Wherever a transition metal gives rise to the two series of salt, the lower one is an ionic compounds while the higher one is a covalent compound.

It can be seen that second ionization energy of the first transition series increases almost regularly from Sc to Zn. Values for Cr and Cu are, however, sufficiently higher than those of their neighbours on the left as well as right. It is due to the stable configuration of Cr+ (3d5-exactly half filled) and Cu+ (3d10. completely filled).

Removal of one electron means they change from a more stable to less stable configuration and is therefore difficult.

Table 11 : Ionization Energies of the Elements of 3RD Series

	Ionization energy (KJ mol^{-1})		
3d Element	*First*	*Second*	*Third*
Sc	632	1245	2590
Ti	650	1320	2721
V	659	1376	2873
Cr	652	1635	2994
Mn	716	1513	3258
Fe	762	1564	2963
Co	758	1647	3237
Ni	736	1756	3400
Cu	744	1961	3560
Zn	906	1736	3838

Alloys

Transition metals are know to form alloys with each other . The reason is that they have almost similar sizes and atoms of one metal can easily take up positions in the crystal lattice of the other. Thus, we are getting alloys of Cr—Ni, Cr— Ni—Fe, Cr—V—Fe, Mn—Fe, etc. Alloys are generally harder and higher melting than the parent metals. They have been more resistant to corrosion than their constituents.

Interstitial Compounds

Interstitial compounds are compounds of indefinite structure and composition. These are formed by the incorporation of small non-metal atoms in the spaces between the metal atoms, *i.e.,* the interstices of the lattice. Examples are $Fe_{0.94}O$, $Fe_{0.81}O$, $VSe_{0.98}$ and $VSe_{1.2}$.

Transition metals form interstitial compounds particularly with the *small* non-metallic atoms hydrogen, boron, carbon and nitrogen. The atoms enter into the vacant holes (interstitial sites) between the packed atoms of the crystalline metal. Interstitial compounds have variable composition, i.e, they are non-stoichiometric. For examples : $TiH_{1.7}$, $VH_{0.56}$, etc.

Nontoichiometry is shown particularly among transition metal compounds of the Group VI elements (O, S,Se. Te). It is mostly due to the variable valency of transition elements.

For example, copper is precipitated from a solution containing Cu^{2+} by passing in H_2S. The sulphide is completely insoluble, but this is not used as a gravimetric method for analysing for Cu because the precipitate is a mixture of CuS and Cu_2S. Sometimes nonstoichiometry is caused by defects in the solid structures.

Interstitial compounds of transition metals, *viz,* hydrides, borides, carbides and nitrides are usually prepared by heating the metal with hydrogen, boron, carbon (or hydrocarbon), and nitrogen (or ammonia.) respectively.

Interstitial compounds have the properties of alloys. They are extremely hard, possess good electrical conductivity and are high melting points (titanium carbide melts at 3683° K, titanium nitride melts at 3473°K).

Abundance

Three of the transition metals are very abundant in the earth's crust. Fe is the fourth most abundant element by weight, Ti the ninth and Mn the twelth.

The first row of transition elements largely follow Harkins' rule that elements with an even atomic number are in general more abundant than their neighbours with odd atomic number. Manganese in an exception.

The second and third row elements are much less abundant than the first row. Tc does not occur in nature. Of the last six elements in the second and third rows (Tc, Ru, Rh, Pd, Ag, Cd; Re, Os, Ir, Pr, Au, Hg) none occurs to an extent of more than 0.16 parts per million (ppm) in the earth's crust.

There are however, two generalizations that apply equally well to both the transition and non-transition elements :

(i) High oxidation states are usually stabilized in alkaline media.

(ii) The capacity of the halogens to bring out high oxidation states varies in the order of F > Cl > Br > I.

COMPARISON OF THE ELEMENTS OF SECOND AND THIRD TRANSITION SERIES WITH THE FIRST TRANSITION SERIES

In a given group, elements of the second and third series have been found to resemble each other but different from that of the first series. Some of such important points are described as follows :

1. Size

The second row elements are all larger than the first row elements. Because of the lanthanide contraction the radii of the third row are almost the same as those for the second row.

2. Metal-metal Bonding and Cluster Compounds

Metal-metal (M–M) bonding occurs not only in the metal themselves, but also in some compounds, M–M bonding is quite rare in the first row transition elements. It occurs only in a few carbonyl compounds such as $Mn_2(CO)_{10}$, $Fe_2(CO)_9$, $CO_2(CO)_8$, $Fe_3(CO)_{12}$ and $Co_4(CO)_{12}$ and in carboxylate complexes such as chromium (II) acetate $Cr_2(CH_3COO)_4(H_2O)_2$, and in solid nickel dimethylglyoxime.

In the second and third row elements M–M bonds are much more common :

(i) They form carbonyls with M-M bonds similar to those from the first row such as $Ru_3(CO)_{12}$, $Os_3(CO)_{12}$, $RH_4(CO)_{12}$ and $Ir_4(CO)_{12}$ and a type not formed by the first row $Rh_6(CO)_{16}$.

(ii) The metals Mo, Ru and Rh form binuclear carboxylate complexes such as $Mo_2(CH_3COO)_4(H_2O)_2$ which are similar to chromium (II) acetate.

(iii) The halide ions $[Re_2Cl_8]^{2-}$ and $[Mo_2Cl_9]^{3-}$ also have M–M bonds.

(iv) The lower halides of several elements have a group of three or six metal atoms bonded together and are called *cluster compounds*. The elements are :

Nb	Mo	
Ta	W	Re

$[Nb_6Cl_{12}]^{2+}$ and $[Ta_6Cl_{12}]^{2+}$ have unusual structures. Both contain six metal atoms arranged in a cluster at the corners of an octahedron, with 12 bridging halogen atoms across the corners. There is extensive M–M bonding within the octahedron. The so-called 'dihalides' of Mo and W are really Mo_6Cl_{12} and W_6Br_{12}, and those contain the $[M_6X_8]^{4+}$ ion. This too has remarkable structure. Six metal atoms are arranged in a cluster at the corners of an octahedron, with eight halogen atoms located above each of the eight face of the octahedron and 'bonded' to three metal atoms. $ReCl_3$ is really trimeric Re_3Cl_9, and comprises a triangle of three Re atoms with three bridging halogen atoms across the three corners, and six halogen atoms that bridge to other Re_3Cl_9 units.

3. Stability of Oxidation States

The (+II) and (+III) states are important for all the first row transitition elements. Simple ions M^{2+} and M^{3+} are common with the first row but are less important for second and third row elements, which have few ionic compounds. Similarly the first row form a large number of extremely stable complexes such as $[Cr^{III}Cl_6]^{3-}$ and $[CO^{III}(NH_3)_6]^{3+}$. No equivalent complexes of Mo or W, or Rh or Ir, are known.

The higher oxidation states of the second and third row elements are more important and much more stable than those of the first row elements. Thus the chromate ion $[CrO_4]^{2-}$ is strong oxidizing agent but molybdate $[MoO_4]^{2-}$ and tungstate $[WO_4]^{2-}$ are stable. Similarly the permanganate ion $[MnO_4]^-$ is a strong oxidizing agent but pertechnate $[TcO_4]^-$ and perrhenate $[ReO_4]^-$ ions are stable. Some compounds exist in high oxidation states which have no counter parts in the first row, for example WCl_6, ReF_7, OsO_4 and PtF_6.

4. Complex

First series transition elements are able to form more stable complexes with N,O and F donors, while those of second and third series are able to from more stable complexes with P, S and X donors.

5. Coordination Number

The coordination number 6 is widespread in the transition elements, giving an octahedral structure. The coordination number 4 is much less common, giving tetrahedral and square planar complexes. Coordination numbers of 7 and 8 are uncommon for the first row but are much more common in the early members of the second and third rows. Thus in $Na_3[ZrF_7]$ the $[ZrF_7]^{3-}$ is a pentagonal bipyramid, and in $(NH_4)_3[ZrF_7]$

it is capped trigonal prism. In $Cu_2[ZrF_8]$ the Zr is at the centre of a square antiprism.

The four coordinate complexes of the first series of transition elements are mainly tetrahedral, whereas that of second and third series are usually square planar.

6. Magnestism

When transition elements form ochtahedral complexes, the d levels are split into t_{2g} and e_g sub-levels. Consider a first row element. If the ligands possess a strong field they cause a large difference in energy between these two sub-levels. Thus the electrons occupying the d level fill the lower t_{2g} level even if this means they must be paired. If pairing of electrons occurs, the complex is called low *spin or spin paired.* Alternatively if the ligands have only a weak field the splitting is small, and only when each of the t_{2g} and e_g levels contains one electron does pairing of spins occur. Such complexes are called *high spin* or *spin free.* Thus with a first row element the strength of the ligand field determines whether a low-spin or a high-spin complex is formed.

The second and third row transition elements tend to give low-spin complexes, that is, it is more favourable in terms of energy to pair electrons in the lower energy d levels rather than use the higher levels, regardless of the ligand.

The spin only formula gives reasonable agreement relating the observed magnetic moment of first row transition metal complexes to the number of unpaired electrons. For the second and third row transition elements the orbital contribution is significant, and in addition spin orbit coupling may occur. Thus the spin only approximation is no longer valid, and more complicated equation must be used. Thus the simple interpretation of magnetic moments in terms of the number of unpaired electrons cannot be extended from the first row of transition elements to the second and third rows. The second and third rows also show extensive temperature dependent paramagnetism. This is explained by the spin orbit coupling removing the from the lowest energy level in the ground state.

7. Isopoly and Heteropoly Anions

The second and third series elements are having more tendency to form the isopoly and heteropoly anions than the first series elements.

8. Abundance

The ten first row transition elements are reasonably common and make up 6.79% of the earth's crust. The remaining transition elements are mostly very scarce. Even though the abundance of Zr is 162 ppm, La 31 ppm, Y 31 ppm and Nb 20 ppm, the 20 elements in the second and third rows elements together make up only 0.025% of the earth's crust. Tc does not occur in nature.

THALLIUM

Crookes in 1861 examined spectroscopically the residues of selenium from a sulphuric acid factory in Hartz Mountains. He expected to find tellurium, as he observed a bright green line in the spectrum of the flue dust. This line did not belong to any of the known element of the known element of the day. Crookes on further investigations proved it to be due to a new element of aluminium group. He named this new element *thallium*. In 1862 *Lamy* obtained its large quantity from lead chamber deposits of a sulphuric acid plant.

Occurrence

Thallium is not found native. Some of the thallium containing minerals are:

1. Crookesite : $(Cu, TI, Ag)_2$ Se
2. Lorandite : $(TIAsS_2)$
3. Huchinsonite : $(TI, Ag, Cu)_2S. PbS._2As_2S_3)$
4. Urbaite : $(TIAs_2SbS_5.)$

The most fertile sources are the flue dust of iron furances and sulphuric acid plants using thalli-ferrous ores. Some of the flue dusts contain as much as 8 per cent thallium.

Extraction

The finely powdered ore is treated with aqua regia. After the addition of H_2SO_4, it is evaporated to remove excess of acid. The solution is then diluted and neutralised with Na_2CO_3, Potassium cyanide is added to precipitate lead and bismuth which are filtered off. The filtrate is then treated with hydrogen sulphide, which precipitates thallium, cadmium and mercury. Very dil. H_2SO_4 dissolves the thallium sulphide only, leaving behind insoluble cadmium and mercury sulphides.

Process

Flue dust from sulphuric acid plant having thallium in the form of Tl_2O and the oxides of associated elements such as Cu, Pb, Bi, As and Fe is dissolved in aqua-regia and the solution is evaporated ti drive away the nitric acid. After whole of the nitric acid is removed, hydrogen sulphide gas is allowed to pass in the solution to precipitate the sulphides of Cu, Pb, Bi and As. The filtrate is next treated with ammonia solution so as to precipitate iron as hydroxide. The clear filtrate obtained after removing these associated elements is treated with potassium iodide to precipitate thallium as thallous iodide, TlI. Thallous iodide formed is next reduced with zinc sulphuric acid or with metal sodium when the metal thallium is obtained.

$$TlOH + KI \rightarrow TlI + KOH$$

$$TlI + H \rightarrow Tl + HI$$

or $$TlI + Na \rightarrow Tl + NaI$$

Alternatively, thallous chloride precipitate can be reduced by fusion with sodium carbonate and potassium cyanide to obtain the metal. In another method chloride is also sometimes converted to thallous sulphate, Tl_2SO_4 by treatment with hot concentrated sulphuric acid. After crystallisation, thallium is obtained by treating with zinc or by electrolysis.

$$2TlCl + H_2SO_4 \rightarrow Tl_2SO_4 + 2HCl$$

$$Tl_2SO_4 + Zn \rightarrow 2Tl + ZnSO_4$$

Flow sheet for the extraction of Thallium from Flue dust on next page 38 :

From the Residues of Lead Chamber in the Manufacture of Sulphuric Acid

Lamy's Process : The mass is first of all treated with lead oxide or lime and then extracted with sufficient hot water several times. The extract which is obtained on filtration is concentrated to a great extent. In the concentrated solution, hydrochloric acid is added which precipitates out thallium as thallous chloride. The thallous chloride is next decomposed with hot sulphuric acid, the thallous hydrogen sulphate, $TlHSO_4$ formed, is dissolved in water and treated with hydrogen sulphide gas, when lead, silver and mercury are precipitated whereas thallium remains in the solution. The solution is now concentrated by evaporation and thallium gets precipitated by adding metal zinc or also by electrolysis.

Purification of Thallium

(a) The sulphuric acid solution from the flue dust is treated with alkaline solution of potassium iodide which precipitates out TII. The iodide is then treated with Na_2S to give thallous sulphide which is again converted into sulphate by H_2SO_4. From this sulphate solution the metal is precipitated by adding zinc or by electrolysis.

(b) Crude thallium material is transformed into sulphate and dilute solution is acidified. H_2S is then passed to precipitate Hg, Cu, Ag, As, Sb and Bi. Ammonia is added to remove iron and aluminium. The purified thallous sulphate is concentrated to crystallisastion.

Metallurgy : Thallium is very easily prepared from its compounds in various ways :

1. Fusion of the thallous chloride or iodide with Na_2CO_3 and KCN.]

$$2TlCl + Na_2CO_3 + 2KCN \rightarrow 2Tl + 2NaCl + K_2CO_3 + (CN)_2$$

2. By adding zinc to thallous sulphate or chloride

$$Zn + Tl_2SO_4 \rightarrow ZnSo_4 + 2Tl$$

3. By heating Tl_2CO_3 with carbon

$$Ti_2CO_3 + 2C \rightarrow 2Tl + 2CO$$

4. Thallium may be produced by the electrolysis of oxide of thallium with B_2O_3. The temperature required is 900°C. Graphite crucible is used as cathode and graphite or gas carbon rod as anode. The metal deposited is free from alkaline earth metals.

5. Recently the electrolysis of thallous formate has been found to give spectroscopically pure thallium.

Properties Physical

1. Thallium is a white metal with blue tint. The freshly cut surface has a brilliant metallic lustre.

2. Pure thallium is almost as soft as sodium and may easily be cut with a knife or scratched with finger nails.

3. It leaves a black streak on paper (similar to lead).
4. It is a malleable but has little tenacity.
5. It has a crystalline structure and emits a sound similar to tin cry.
6. It melts at 302°C and boils at 1650°C
7. It exists in two modification. The transition temperature being 232°C.

$$A \xrightarrow{232°C} B$$

Chemical Properties

1. **Reaction with oxygen :** Pure thallium oxidises rather quickly in air and in water containing dissolved air. Thallous hydroxide TiOH is formed during oxidation in the presence of water.
2. **Reaction with water :** Thallium decomposes water at red heat. It does not react with oxygen free water at ordinary temperature and so the metal is preserved in water free from oxygen.
3. **Reaction with halogen :** It combines directly with halogens forming halides.
4. **Reaction with S, Se, Te, P, and Sb :** It does not react with H, C, Si,or B.
5. **Reaction with dil. HNO_3 and H_2SO_4 :** It dissolves in dilute acids evolving hydrogen.

$$2Tl + 2HNO_3 \rightarrow 2TlNO_3 + H_2$$

$$2Ti + H_2SO_4 \rightarrow Tl_2SO_4 + H_2$$

6. **Reaction with HCl :** It is very slightly affected by HCl due to formation of insoluble TlCl.
7. **Reaction with metals :** Thallium alloys with many metals especially potassium, lead, mercury, magnesium, calcium, gold, silver, cadmium, copper and zinc.

Uses

1. **As alloys :** Thallium may be used of the preparation of alloys. *Thallium*-Pb alloys are resistant to ordinary atmosphere and anodic corrosion. *Thallium alloys with tin and indium* show super electrical conductivity below that of liquid air.

2. **Medicines :** Thallous acetate has found use in medicines for the treatment of certain kinds of tuberculosis and skin diseases as ring worm.
3. **Photography :** Thallous chloride and iodide may be used in photography for producing various shades, yellow, green, etc.
4. **Catalyst :** Thallium chloride is used as a catalyst in chlorination. Asbestos fibres soaked in solution of Tl_2SO_4 are used as catalyst for hydrogenation of nitrobenzene.
5. **Glass industry :** Thallium compounds (Tl_2O) due to high refracting power are used in the manufacture of certain kinds of optical glass.
6. **In analytical chemistry :** *Thallium carbonate* is an excellent standard for titration with acid or potassium iodate. *Thallous nitrate* is useful in determining iodine in presence of chlorine. It is also useful in determining thorium, uranium, gold and platinum.

Compounds of Thallium : Thallium forms compounds in the oxidation states of + 1 and + 3. The compounds containing monovalent thallium are very stable and are the common compounds of thallium. The salts of trivalent thallium are easily reduced to those of monovalent thallium, so trivalent thallium is because of high positive electrode potential for the reaction.

$$Tl^{3+} \text{ (aq.)} + 2e^- \rightarrow Ti^+ \text{ (aq.)} : E^\circ = +\ 1.25 \text{ volts.}$$

Monovalent state is important for thallium. The thallous salts such as TlCl, TlBr, TlI, Tl_2S and Tl_2CrO_4 are sparingly soluble like silver salts. Monovalent thallium resembles alkali metals in forming colourless crystalline salts. Thallous salts of weak acids are weakly basic in aqueous solution because of hydrolysis. Like the alkali metal hydroxides and carbonates thallous hydroxide and carbonate are quite soluble in water and basic in character.

The electrode potential for the change $Ti^+ + e^- \rightarrow Tl$ is —0.336 volt.

General Properties of Thallous Salts

1. Thallous chloride is precipitated as white precipitate as white precipitate from aqueous solution of thallous salt on the addition of dilute HCl or solution of a soluble choloride. Thallous iodide

is obtained as a yellow precipitate when potassium iodide is added to Ti(l) solutions.

Thallous iodide is only sparingly soluble in $Na_2S_2O_3$ solution and thus can be distinguished from real iodide. Thallous chromate (yellow) is precipitated from thallous salt solution on addition of K_2CrO_4.

2. When H_2S is passed through a solution of Ti(l) sàlt acidified with H_2SO_4 and HNO_3, thallous sulphide is not precipitated. It is incompletely precipitated from neutral solution by H_2S. Complete precipitation of thallous sulphide takes place if thallous slt solution is acidified with dilute acetic acid. The precipitate of thallous sulphide is insoluble in bases.
3. Thallous salts do not precipitate with NH_4OH or alkalies solution because thallous hydroxide readily dissolves.
4. Alkali metal carbonates do not precipitate thallous carbonate from dilute solution of thallous salts because thallous carbonate dissolves in water. It can be precipitated only from concentrated solution.
5. Soluble thallous salts yield a white crystalline precipitate of TlCNS When KCNS is added to the salt solution. This crystalline precipitate is soluble in boiling water.

General Properties of Thallic Salts

1. Thallic salts are unstable and are fairly strong oxidising agents.
2. Aqueous solution of thallic salts such as of thallic nitrate and thallic sulphate are hydrolysed on boiling and produce the precipitate of thallic hydroxide.
3. With potassium iodide the thallic salts produce a mixture of thallous and thallic iodide.
4. H_2S does not precipitate thallium, sulphide when it is passed through the solution of thallic salts acidified with HNO_3 or H_2SO_4. In this respect thallous and thallic salts are similar in their behaviour. Thallic salts yield precipitate of thallous sulphide when H_2S is passed through neutral or alkaline solution of Tl (III) salt. Tl_2S is also precipitated from thallic salts if the solution has been acidified with dilute acetic acid.

5. When aqueous solution of NH_4OH, KOH or NaOH are added to thallic salt solution, brown coloured thallic hydroxide is precipitated. When this precipitate is exposed to air, it slowly changes into TlO (OH) which is almost insoluble in alkali solutions and dissolves only slowly in acids.

Oxides of Thallium

Thallium forms number of oxides but the two important oxides are thallous oxide, Tl_2O and thallic oxide, Tl_2O_3.

THALLOUS OXIDE, Tl_2O

Preparation

It can be prepared by heating thallium in air or by heating thallous hydroxide to 100°C in absence of air.

Properties

1. It is a black powder which melts at 300°C into a dark yellow liquid.
2. When heated to redness in hydrogen or in carbon monoxide, it is reduced to metal.
3. It is readily attacked by chlorine.

$$2Tl_2O + H_2SO_4 \rightarrow 4TlCl + O_2$$

4. It reacts with acids to from thallous salts.

$$Tl_2O + H_2SO_4 \rightarrow Tl_2SO_4 + H_2O$$

THALLIC OXIDE Tl_2O_3

Preparation

1. Thallic oxide is obtained when the metal in molten condition reacts with oxygen.
2. It can also be prepared by the electrolysis of thallic sulphate or nitrate using platinum electrodes.

Properties

1. It is black or brown powder. On heating to 800 –1000°C, it decomposes to give oxygen.

2. It is reduced to metal and thallous oxide by hydrogen or carbon mono-oxide at red heat.

Chlorides of Thallium

Thallous Chloride TlCl

Preparation

1. It can be prepared by treating thallous salts with hydrochloric acid or a soluble chloride.
2. It can be also prepared by reducing thallic chloride with SO_2 or $SnCl_2$.

Properties

1. It is a white substance, sparingly soluble in water.
2. It melts at 430°C and boils at 720°C.
3. Thallous chloride on heating in dry chlorine forms thallium dichloride, ($TlCl_2$), but suspended in water and treated with chlorine, it forms $TlCl_3$.

Uses : It is used in photography for producing various shades like yellow, green etc.

Thallic Chloride $TlCl_3$

Preparation

1. It is prepared by passing chlorine into an aqueous suspension of thallous chloride.

$$TlCl + Cl_2 \rightarrow TlCl_3$$

2. It can also be prepared by heating thallous chloride with aqua regia.

Properties

1. It is colourless crystalline, deliquescent substance which melts at 45°C.
2. Hydrated thallic chloride easily dissolves in alcohol and ether forming $TlCl_2 . C_2H_5OH$ and $TlCl_3 . (C_2H_5)_2O$ respectively.
3. It combines with ammonia to form $TlCl_3.3NH_3$ and also reacts with pyridine.

4. It also forms double salts with salts such as NH_4Cl and LiCl, etc.

Uses : It is used in tungsten lamps for increasing their life (by liberating chlorine).

Thallous Sulphate

Thallous sulphate, Tl_2SO_4 : It may be prepared by dissolving metallic thallium, its oxide, hydroxide or carbonate in dilute sulphuric acid.

$$2Tl + H_2SO_4 \rightarrow Tl_2SO_4 + H_2$$

It forms needle shaped colourless crystals. It is isomorphous with potassium, rubidium and caesium sulphates. It forms alums with other metal sulphate, *eg.*, Mg, Zn, Fe, etc. When oxidised with alkaline hydrogen peroxide or potassium ferricyanide, it forms thallic sulphate. It is mainly used for the preparation of other salts.

$$Tl_2SO_4 + H_2S \rightarrow Tl_2S + H_2SO_4$$

$$Tl_2SO_4 + 2KX \rightarrow 2TlX + K_2SO_4$$

$$Tl_2SO_4 + Ba(OH)_2 \rightarrow 2Tl(OH) + BaSO_4$$

Thallic sulphate, $Tl_2(SO_4)\ 7H_2O$: It may be prepared either by treating thallic oxide with dilute sulphuric acid or by oxidising thallous sulphate with peroxides.

$$Tl_2O_3 + 3H_2SO_4 \rightarrow Tl_2(SO_4)_3 + 3H_2O$$

It is decomposed in aqueous solution, forming a basic salt $Tl(OH)SO_4.2H_2O$. With potassium sulphate, it forms $K_2SO_4.Tl_2(SO_4)_3.$ $8H_2O$ which is not a true alum.

Estimation

Thallium can be gravimetrically estimated either as thallous iodide or as chromate.

(a) Thallous salt is precipitated as thallous iodide by the addition of potassium iodide solution at 80°C. The yellow coloured precipitate obtained is washed with alcohol, dried at 170°C and then weighed.

(b) Thallous chromate can be precipitated by the addition of 10% potassium chromate to a thallous salt solution heated at 80°C. This method is applicable to thallous salts only and, therefore thallic salts are first reduced to thallous state by passing sulphur dioxide gas into thallic salt solution.

The precipitate is washed with 1% K_2CrO_4 solution and finally with 50% alcohol. It is dried at 120°C and weighed.

BERYLLIUM

General Considerations

Beryllium along with it congeners, magnesium, calcium, strontium, barium and radium belong to IIA group of the periodic table. It is a bridge-element of its group and thus it resembles members of its own group and typical element, aluminium, of III group.

Beryllium has $[He]_2S^2$ electronic configuration, has the highest electronegativity (1.5), highest ionisation potential (18.2 e.v.) and smallest ionic (Be^{2+}, 0.31Å) radius in the group. All these values of beryllium make it a most suitable element for exhibiting covalent character and highest, complexing ability in the group.

The chemistry of beryllium is dominated by a tetrahedral geometry and to a lesser extent by a linear co-ordination. Beryllium is member of the first short period and hence can have not more than four orbitals (three p's and one s) in the n = 2 quantum shell.

Such a quantum shell will not have a d orbital. Therefore, beryllium can have a maximum of four co-ordination and thus can acquire the next inert gas configuration through complex formation. Four coordination can be attained either by tetrahedral (Sp_3) hybridisation or a square planar (dsp^2) geometry. The dsp^2 type of geometry is energetically not favoured because the only d-orbital possible with beryllium has much high energy (3d).

The possibility of favoured co-ordination in the element is unfavourable because in that case we will have to invoke an sp-hybridisation giving two sp hybrid orbitals at 180° to each other. But such a co-ordination is only provided by a valency shell of only 4(= 2+2) elements around the metal. For this beryllium will have to raise its co-ordination number to 4 providing it inert gas configuration.

Because of this tendency beryllium compounds like $BeCl_2$ with an apparent co-ordination number 2 are really polymeric with a coordination number 4.

COMPARISON OF BERYLLIUM WITH ALUMINIUM : DIAGONAL RELATIONSHIP

The charge/radius ratio in case of beryllium and aluminium is almost the same and hence these two elements possess similar deforming power. This ratio is far closed for Be- Al to each other than the lithium-magnesium pair. As result of this the properties of Be/Al are much more close than those of Li/Mg. Indeed Be/Al pair is considered as the best pair in the periodic table:

	Be	Al
Ionic radius (Å)	0.31	0.51
Charge on the ion	+2	+3
Charge/radius	6.4	6.3

The point of similarities are :

1. Be and al exhibit strong tendency to covalence and complex formation. The halides of aluminium are covalent and dimeric while beryllium halides are covalent and polymeric.
2. In aqueous solution, the aquo ions are formed :

$$[Be(H_2O)_4]^{2+} \text{ and } [Al(H_2O)_6]^{3+}$$

3. Both Be $(OH)_2$ and $Al(OH)_3$ have amphoteric properties.
4. The standard potential of Be is – 1.85 volt and that of Al is – 1.66 volt so that the metals should readily dissolve in acids to form metal salts.

History

Vauquelin in 1798 observed an unknown element during the chemical analysis of the mineral beryl. *Wohler in 1928* prepared beryllium for the first time by reducing beryllium chloride with potassium. Pure beryllium was prepared by *Lebeau in 1898* by the electrolysis of sodium beryllium fluoride *Goldschmidt in 1832* devised the first commercial process by using a mixture of fluorides of beryllium and barium.

Occurrence

Beryllium in never found native. It occurs in a number of minerals. Most of which are silicate ores. Some of the important minerals of this element are.

1. Beryl	$3BeO.Al_2O_3.6SiO_2$	(Be = 14%)
2. Phenacite	$2BeO.SiO_2$ or $Be_2[SiO_4]$	(Be = 16.9%)
3. Bromellite	BeO	(Be = 36%)
4. Bertrandite	$4BeO.2SiO_2.H_2O$	(Be = 15.2%)

Extraction

Extraction of beryllium is generally done from beryl.

1. French Method

Finely ground beryl is mixed with sodium silico fluoride and heated to 650°C. The double fluoride, which is soluble in water. $[2NaF.BeF_2]$ is formed. Aluminium is mineral is converted into insoluble Na_3AlF_6.

$$2BeO + Na2SiF_6 \rightarrow 2(BeF_2.NaF) + SiO_2$$

Silica remains insoluble. The fused mass is extracted with boiling water which dissolves beryllium salt. The filtrate contains sodium beryllium fluoride and small amounts of aluminium and iron salts. This solution is added to large amounts of $(NH_4)_2CO_3$ with the result $Al(OH)_3$ is precipitated. The filtrate is now boiled to precipitate out iron as its basic carbonate while the last traces of iron are eliminated by bubbling H_2S. It is then filtered and from filtrate the basic beryllium carbonate is precipitated by boiling it with $(NH_4)_2CO_3$. The precipitate so obtained is heated with HCl to dryness and get solid $BeCl_2$.

2. American Method

In this process finely ground ore is chlorinated at 500°C for several hours. Iron, aluminium and other impurities form volatile chlorides and are removed. The temperature is then raised to about 900°C and $BeCl_2$ which is volatile at this temperature condenses into the condenser. The residue left mostly contains silica. The beryllium chloride produced is electrolysed in a fused mixture of beryllium chloride and sodium chloride.

3. German Method

Finely powdered mineral is fused with Na_2SiF_6 for several hours at 85°C. Na_2BeF_4 is extracted with water. Impurities of iron are precipitated by bubbling air through solution. To the filtrate KOH is added to precipitate $Be(OH)_2$ with some CaF_2.

The precipitate is treated with HF. Calcium fluoride which remains insoluble is filtered off. The filtrate is evaporated which yields crystals of beryllium oxyfluoride ($5BeF_2.BeO$).

4. Alkali Fusion Method

The finely ground beryl is fused with twice of its weight of KOH or K_2CO_3. The melt is cooled and ground to a fine powder. The powder is covered with water and then treated with an excess of H_2SO_4. It is then heated. Heating is continued with stirring till a fine white powder is obtained. The residue is then dissolve in water and filtered to remove SiO_2.

$$3BeO.Al_2O_3.6SiO_2 + 6k_2CO_3 \rightarrow 3BeO + Al_2O_3 + 6k_2SiO_3 + 6CO_2\uparrow$$

$$3[BeO + H_2SO_4 \rightarrow BeSo_4 + H_2O]$$

$$Al_2O_3 + 3H_2SO_4 \rightarrow Al_2(SO_4)_3 + 3H_2O$$

$$6[K_2SiO_3 + H_2SO_4 \rightarrow K_2SO_4 + SiO_2 + H_2O]$$

$$3BeO.Al_2O_3.6SiO_2 + 6k_2CO_3 + 12H_2SO_4 = 3BeSO_4 + Al_2(SO_4)_3 + 6K_2SO_4 + 6SiO_2 + 12H_2O + 6CO_2$$

The filtrate contains sulphates of beryllium, aluminium and potassium. It is carefully concentrated so that aluminium crystallises out as potash alum $K_2SO_4.Al_2(SO_4)_3.24H_2O$. The mother liquor contains $BeSO_4$ and traces of K_2SO_4 and $Al_2(SO_4)_3$ and some other impurities. From this $BeSO_4$ is separated by any one of the following methods.

(a) Methylamine or ethylamine is added to the solution containing $BeSO_4$ so that only $Be(OH)_2$ is precipitated. This is filtered, dried and ignited to oxide.

(b) Acetic acid is added to the solution and the whole solution is boiled so that basic beryllium acetate is formed. The solution is shaken with chloroform so that basic beryllium acetate passes out into chloroform but acetates of iron and aluminium remain in the aqueous layer. The chloroform layer is separated and evaporated. The solid basic beryllium acetate is lgnited to oxide.

Preparation of the Metal

Two different methods are in use for this purpose.

(i) **Stock and Goldschmidt's Electrolytic Method :** In this method beryllium fluoride is dissolved in a fused mixture of sodium and barium fluorides. The resulting complex salt Na_2BaBeF_6 is then made to electrolyse at 1400° C in a graphite crucible, which acts as the anode.

The cathode is generally a water cooled rotating iron tube on which tough rods of beryllium are formed. When the temperature reaches to about 700°C, beryllium oxyfluoride and sodium fluoride are added into the electrolyte so as to raise the fusion point and thereby keep the liquid from volatilizing. The temperature is now raised to 1300–1400°C. Beryllium is deposited at the rotating cathode and is of 99.8 percent purity.

In America, a mixture of beryllium chloride and sodium chloride is used for electrolysis. This mixture is taken in a ferro chrome vessel and is kept at 820°C. The electrolysis is done using nickel cathode and graphite anode in an atmosphere of Argon.

(ii) Reduction of the oxide method : This method is used by the Brush Beryllium Company. It involves the reduction of beryllium oxide by carbon.

Beryllium, oxide is powdered and mixed with powdered coke (copper or any other alloying metal can be added in order to get the alloy of beryllium). This mixture into an electric are furnace. The lower part of the furnace is made of carbon, and serves as the receptacle of the charge and also as one of the electrodes.

Metallurgy

1. Wohler prepared metallic beryllium by heating anhydrous $BeCl_2$ with metallic potassium.

$$BeCl_2 + 2K \rightarrow Be + 2KCl$$

2. By heating double fluorides of potassium and beryllium, $2KF.BeF_2$ with potassium, when beryllium metal is obtained.

3. Sodium and potassium florides are added to pure BeF_2 and the mixture is placed in a nickel crucible. The double salt after melting is electronlysed passing a current of 20 amperes at 80 volts.

   Graphite rod serves as anode and nickel crucible as cathode. In a few hours the crystals of beryllium deposit in the crucible which are washed with water absolute alcohol.

4. Beryllium metal is also obtained by electrolysis of fluorides of beryllium and barium. The molten-beryllium separates out at the water cooled iron cathode.

Purification

Beryllium of highest purity can be obtained by subliming the crude metal in vacuum at about 2000°C.

GENERAL CHARACTERISTICS OF D-BLOCK ELEMENTS

The elements having partly filled d sub-shell in their elementary form or in their chemically significant oxidation states are called *d-block elements.*

The d-block elements are also termed as *transition elements*. These have been so called because their properties have been intermediate between those of the s and p-block elements.

They make up three complete rows of 10 elements and an incomplete fourth row in the periodic table as shown below :

IIIB	*IVB*	*VB*	*VIB*	*VIIB*		*VIII*		*IB*	*IIB*
Sc $3d^1 4s^2$	Ti	V	Cr	Mn	Fe	Co	Ni	Cu	Zn $3d^{10} 4s^2$
Y $4d^1, 5s^2$	Zr	Nb	Mo	Tc	Ru	Rh	Pd	Ag	Cd $4d^{10}, 5s^2$
La $5d^1, 6s^2$	Hf	Ta	W	Re	Os	Ir	Pt	Au	Hg $5d^{10} 6s^2$
Ac $6d^1, 7s^2$	Ku	Ha							

From the above table, it is evident that there are four series of d-block elements each having ten elements;

First Series

It is also called 3d-series. It includes elements from scandium to zinc (at no. 21-30).

Second Series

It is also called 4d-series. It includes elements from yttrium to cadmium (at. no. 39-48).

Third Series

It is also called 5d-series. It includes elements from lanthanum (at. no. 72) to mercury (at no. 80).

Fourth Series

It is incomplete 6d-series in which the 6d-orbitals start filling. It consists of only three elements viz, Ac (At. No. 89), Ku (104) and Ha (105). The position of the incomplete fourth series is discussed along with the f-block elements.

As zinc, cadmium and mercury are having a completely filled (n–1) d subshell in their elementary form as well as in the combined state, it means that in according to the definition given above these elements cannot be included in the d-block elements. They are not exhibiting much resemblance with other d-block elements except for their ability to form complexes with ligands like NH_3, amines and halide ions, etc. However, they are usually studied with the d-block elements so as to maintain a rational classification of elements.

Further *elements of group IIB* (Sc,Y, La and Ac) have been found to be different *markedly in their properties from that of other transition elements, viz.,* their compounds are uniformly trivalent, diamagnetic and colourless whereas the typical transition elements show variable oxidation states, form coloured ions are paramagnetic. As they are having partly filled d subshell these are regarded as transition elements. Thus, *elements of the group IIIB (Sc, Y, La and Ac) and IIIB (Zn, Cd and Hg) are termed as non-typical transition elements while the other transition elements are termed as typical transition elements.*

In the periodic table, on the left hand side s-block elements are present while on the right-hand side p-block elements are present. In between the two, there are d-block elements. These are also called transition elements because they can be regarded as effecting a transition from the electropositive elements (s-block) to the electronegative elements (p-block)

4

Chemical Analysis of Organic Reagents

INTRODUCTION

Organic reagents in inorganic chemical analysis are those substances which by some form of interaction are used to detect and determine other ions or molecules. Organic reagents have been used for long time in analysis and in fact manifold advantages are derived from them which have been outlined below :

1. They are more specific than conventional reagents under certain conditions. Because of this property they can frequently be used to detect a single type of ion in a complex mixture.
2. They can be more sensitive. Classical semi micro methods of qualitative analysis will usually not detect much less than 100 gammas (1000 γ= 1 mg.) of a substance under specific test solutions. Organic reagents on the other hand are capable of detecting below 10 γ and in certain cases as little as 10^{-3} γ.
3. They can be adopted in quantitative analysis. The organic precipitates are generally insoluble in water at a given pH and can be separated, dried and weighed. Those reactions giving coloured soluble products frequently show a regular change in colour intensity with concentration of the substance under test and so are adoptable to an accurate comparison with colour representing concentration. This is the basic idea in the branch of colourimetry.

Organic reagents are used in various ways in inorganic analysis. Some of the common properties that are used in this way include the following:

1. Formation of a characteristics colour of coloured precipitate
   (a) Organic reagents form direct characteristic colour with the ions to be detected.
   (b) For locating end point, *i.e.,* as indicators employed in volumetric titrations.
2. **Formation of Precipitate due to Difference in Solubility:** This may be made the basis of a gravimetric method.
3. Extraction in organic solvent at controlled pH.
4. **Masking Reagents:** This increases the selectivity of reagents.
5. Difference in volatility.
6. Oxidation or reduction to other valency state.
7. Wash liquids.

The chemical nature of the organic reagent is important in determining the use of above mentioned properties.

Advantages of Organic Precipitating Reagents: There are several ways in which organic reagents are better than inorganic reagents.

1. **Selectivity:** In many bases organic reagents are somewhat specific. For example, cobalt from nickel using u-nitroso-R-naphthol or dimethylglyoxime, aluminium from iron using cupferron, cadmium from copper using quinaldinic acid and thiourea can be precipitated. In some cases, adjustment of the pH will often cause precipitation of one constituent alone. By changing pH and solvent conditions, one constituent after another can sometimes be quickly precipitated out.
2. **Ease of Drying:** Organic reagents produce compounds that have little affinity for water and are dried easily at temperature low enough to prevent decomposition.
3. **Co-precipitation:** Lack of ionic character of organic precipitation products greatly reduces coprecipitation, in most cases. For example the co-precipitation of sodium and potassium with magnesium is much less when the latter is precipitated by 8-hydroxyquinoline than when it is precipitated by phosphate or oxalate.

Some exceptions in metals do occur. It has been little studied but in some cases it is as bad as in inorganic precipitants. Reagents are usually insoluble in water, therefore, it is important not to add an excess of reagent. This is difficult because there is no way to tell when sufficient reagent, without excess, has been added. Any excess that is insoluble would add to the weight of the precipitate.

Disadvantages of Organic Reagents: There are, unfortunately, disadvantages to using organic reagents.

1. **Volatility:** This is the first practical disadvantage of organic reagents. Because chelated compound, are non polar, the forces between individual molecules in the crystal are not very great. Thus, some of the metallic complexes which are precipitated in analytical chemistry are appreciably volatile above 130-150°C and most decompose above this temperature. Therefore, the drying of precipitate should be done below 130°C to avoid decomposition.

2. **Impurities in Reagents:** It is very difficult to prepare an organic reagent of the same degree of purity as say inorganic compounds. Impurities present in organic reagents may enter into the precipitate during the precipitation and cause undesirable reactions. For example,

   (i) Dimethylglyoxime exists in three forms, *i.e.*, α, β and γ. Of these, only the a form is specific for nickel while preparing dimethylglyoxime y-form is also formed along the α-form. If the precipitation is carried out with this impure dimethylglyoxime, then nickel will he precipitated with some other metals because γ-form is specific for nickel whereas γ-form can precipitate other metals also.

   (ii) Keto-enol isomerism in such compounds as dithiozone will cause errors in colorimetric work unless conditions are carefully selected.

3. **Low Solubility of the Reagents in Water:** This is the greatest disadvantage which organic precipitants have in comparison with inorganic precipitants.

4. **High Molecular Weight:** The very high molecular weights of chelate complexes formed with metal ions reduce, proportionately, small errors in precipitation and weighing.

5. **Coloured Compounds:** Chelate compounds are often highly coloured and since they are usually soluble in organic solvents, they lend themselves to colorimetric or photometric methods of analysis.

6. **Solubility in Organic Solvents:** Because of their covalent nature, most metal complexes with organic reagents are soluble in non-polar solvents, chloroform being usually the best. Such solvents are used to extract the complexes for colorimetric determination, or they may be used simply for separating one element from another ; thus the complex of cupferron with tripositive iron is very soluble in ether and chloroform, which allows one to remove large amounts of iron from solutions containing aluminium. In all these extractions, the pH must he carefully adjusted to give the desired separation.

During the last twenty years there has been vast research in this field of chemistry, with the result, today there are over five thousand reagents which could be employed in inorganic analysis. Some of the important reagents have been described below:

1. Dimethylglyoxime: The reagent was discovered by L. Tschngaff (1905) and was applied by 0. Brunck (1907) for the determination of nickel in steel.

$$\begin{array}{c} CH_3\text{—}C=NOH \\ | \\ CH_3\text{—}C=NOH \end{array}$$

This reagent gives bright red precipitate. The precipitation is carried out in ammoniacal medium.

2. $$\begin{array}{c} CH_3\text{—}C=NOH \\ | \\ CH_3\text{—}C=NOH \end{array} + NiCl_2 + 2NH_4OH \rightarrow$$

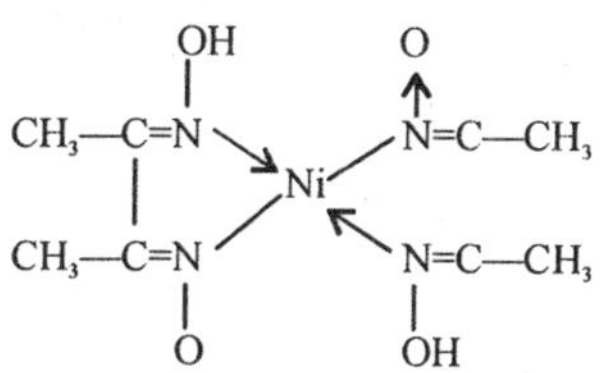

$+ 2NH_4Cl + 2H_2O$

Solution of palladium salts gives characteristic yellow precipitate in dil. HCl or H_2SO_4 but nitrate ions interfere. Solution of bismuth salts also gives a yellow precipitate in presence of EDTA at pH 11.0 to 11.5. For the studies 1% solution of the dimethyl glyoxime in alcohol is employed. In spectrophotometric method, the insoluble precipitate is extracted into chloroform which yields yellow colouration and then the optical density is measured.

2. Thio-urea: Thio-urea forms a yellow precipitate with acid solution of bismuth salts. Coloured complexes are also formed with osmium, selenium, tellurium and mercury.

3. Pyridine: Pyridine forms insoluble complexes with the thiocyanates of Cd, Cu, Ni, Co, Zn and Mn. The insoluble complexes have the general formula $M(SCN)_2\,(C_5H_5N)_n$ when n = 4 for Co and Ni: n = 2 For Cu, Cd and Zn.

Pyridine has also been employed as a solvent in the extraction of insoluble chelates.

4. EDTA: It has been employed in the following analytical methods:

(i) In colorimetric analysis.

(ii) Titrations of metallic ions.

(iii) In the determination of anions.

(iv) As a masking agent.

Detail of the analytical applicability of the reagent has been described in the previous chapter.

5. Rubeanic Acid (Dithoxamide): Rubeanic acid, the diamide of dithio-oxalic acid, presents an equilibrium mixture with its tautometric (aci-) di-imido form:

$$\begin{array}{c} S{=}C-NH_3 \\ | \\ S=C-NH_2 \\ \text{I} \end{array} \rightleftharpoons \begin{array}{c} HSC=NH \\ | \\ HSC=NH \end{array} \rightleftharpoons 2H^+ \left\{ \begin{array}{c} -SC=NH \\ | \\ -SC=NH \\ \text{II} \end{array} \right\}^{-2}$$

The aci-form combines with nickel, cobalt and copper to give coloured water insoluble inner complex salts of the constitution.

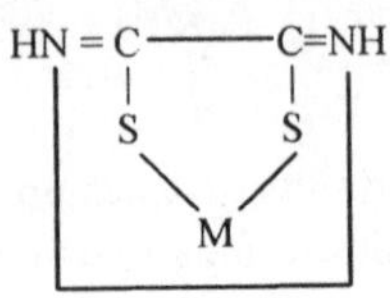

where M = copper (II), cobalt or nickel

The copper, cobalt, and nickel compounds are quantitatively precipitated from strongly ammoniacal solutions.

In strong mineral acid solutions, rubeanic acid reacts only with certain elements of platinum salts.

In addition to above, rubeanic acid forms coloured compounds with silver, mercury (I), bismuth and iron (III).

Rubeanic acid is not only important as spot test reagent but also as a gravimetric ad spectropotometric reagent.

6. Cupron: (a-Benzoin oxime).

C_6H_5—CH—OH
|
C_6H_5—C= N— OH

It is a white crystalline solid. It is sparingly soluble in water but fairly soluble in alcohol. It also exists in isomeric β-form.

Cupron is used as a reagent for the quantitative determination of copper, molybdenum and tungsten. Alcoholic solution of cupron gives a green precipitate with ammonical solution of copper salt. Precipitation of other metals in ammoniacal medium is prevented in the presence of tartaric acid.

The green precipitate is soluble in dilute mineral acids. Copper can be estimated both volumetrically and gravimetrically by this reagent. The green precipitate is an inner complex whose structure is represented as:

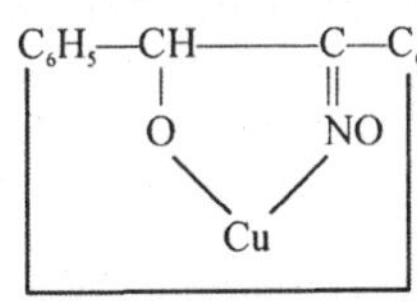

Cupron also precipitates molybdenum and vanadium ions quantitatively but in strongly acid solution vanadium gives a yellow precipitate while molybdenum gives a white precipitate: Tungsten and palladium also appear to be quantitatively precipitated with a-benzoin oxime.

Feigl *et al.*, have recommended the use of the reagent for the detection of basicity in slightly soluble materials such as Al_2O_3, $BaCO_3$ etc.

7. Uranyl Zinc Acetate [$Zn(UO_2)$. $(CH_3COO)_4$]: The reagent is most satisfactory for the identification of sodium. The reagent yields a yellow crystalline precipitate with sodium. The quantitative determination of sodium can be made.

8. Quinaldinic Acid: This organic reagent yields insoluble complexes with Cu, Cd, Zn, Mn, Co, Ni, Pb, Hg, Fe and Pt. It also forms basic salt with metals like Al, Cr, Be and Ti. The reagent which is used as precipitant is 2% aqueous solution of the acid or its sodium salt.

SOLID ELECTROLYTES

In solid electrolytes one or more types of ion diffuses at an appreciable rate. Examples are Ag_2HgI_4 above 50°C and sodium β-alumina where Ag^+ and Na^+, respectively, are mobile.

The low-temperature modification (β) of AgI has the ZnS wurtzite structure and is a poor conductor of electricity; this is generally true for solid salts. At 145.8°C there is asharp transformation to α-AgI, which has a body-centred cubic arrangement of I^-, accompanied by a tremendous increase in ionic conductivity. Such substances are referred to as *solid electrolytes* since their conductivities are closer to those of liquid electrolytes than those of ordinary solid salts.

The Ag^+ ions in solid α-AgI (mp 555°C) can move from their normal sites through positions of lower coordination number. These are interstitial sites (Frenkel defects). At the melting point there is a 12% decrease in conductivity with melting. Other good solid electrolytes, such as $RbAg_4I_5$ and $[(CH_3)_4N]_2Ag_{13}I_{15}$, have passageways of face-centred tetrahedra, fewer than half of which are occupied by Ag^+. For Ag_2HgI_4 at 50.7°C a phase transition occurs, forming the solid electrolyte α-Ag_2HgI_4.

The passageways generally are zig-zag for these solid electrolytes-except for $(C_5H_5NH)Ag_5I_6$ (pyridinium salt), which has an essentially hcp arrangement of I^- (at A and B close-packed positions). The pyridinium ions are located at sites centred about C positions, and all octahedral holes for Ag^+ are lined up along the C directions forming straight channels for motion of Ag^+. This motion is limited by the number of vacancies, but vacancies can be created by movement of Ag^+ ions from octahedral holes into empty tetrahedral holes.

The structure is ordered at –30°, with Ag^+ occupying octahedral sites and only one of the two sets of tetrahedral sites; but the Ag^+

positions are disordered at room temperature and above. Solid electroeytes generally display Frenkel defects; and large highly polarizable anions and highly polarizing cations of low charge promote defects and solid electrolyte behaviour.

Some new high-energy batteries use non-aqueous systems requiring:

(a) High temperature, such as liquid Na-liquid S or liquid Li-metal sulphide with a molten electrolyte,

(b) Solid electrolytes, or

(c) Non- aqueous solvents for the electrolyte for ambient temperature batteries.

One solid electrolyte used is α-alumina, an isomorph of Al_2O_3. Based on x-ray diffraction studies, the empirical formula of β-alumina is not pure Al_2O_3 but instead Na_2O. $11Al_2O_3$. Commonly the β-alumina used contains ~ 25% more Na_2O than indicated by the formula. Beta-alumina has a layer structure featuring four close-packed layers of oxide ions and Al^{3+} in octahedral and tetrahedral sites.

The Na^+ occupy planes between the quadruple oxide layers, these planes contain equal numbers of loosely packed Na^+ and O^{2-} ions. An excess of this ratio of Na^+ ions is accompanied by Al^{3+} vacancies, to maintain charge balance. Adding M_gO or Li_2O creates a stabilized form referred to as β∀-alumina. Both β- and β"-alumina have high ionic conductivity for Na^+ and can be used as solid electrolytes for the liquid Na-liquid S battery.

Any electrochemical cell such as a battery, fuel cell, electrochromic display, or electrochemical sensor, requires an electrolyte. Recently there has been considerable interest in the use of solid electrolytes in these applications, and a commercial device utilizes a solid oxide ion conductor to measure O_2 pressure. Silver tetraiodomercurate (II), Ag_2HgI_4 and β-alumina, a sodium doped version of alumina of composition Na_{1+x} Al_{11} O_{17} + x/2, provide examples of two quite different types of solids in which cations are highly mobile. The former is a soft material with a lattice of low rigidity and the latter is a hard refractory solid. In general, the diffusion rate is proportional to the defect density of the controlling defect.

Below 50°C, Ag_2HgI_4 has an ordered crystal structure in which Ag^+ and Hg^{2+} ions are tetrahedrally coordinated by I^- ions and there are unoccupied tetrahedral holes. At this temperature its electronic and ionic conductivity is low. Above 50°C the Ag^+ and Hg^{2+} cations of Ag^+ HgI_4 are distributed in a disorderly manner over all the tetrahedral sites. At this temperature the material is a good electrical conductor hugely on account of the mobility of the Ag^+ ions.

The weak and easily deformed close packed array of I^- ions gives rise to a low activation energy for ion migration from one lattice site to the next. Solid inorganic electrolytes generally have a low-temperature phase in which the ions are ordered on a subset of lattice sites and at higher temperatures the ions become disordered over the sites and the ionic conductivity increases. There are many similar solid electrolytes having soft lattices, such as AgI and $RbAg_4I_5$, both of which are Ag+ conductors. The mobility of ions in solids can be quite high : the conductivity of $RbAg_4I_5$ at room temperature is greater than that of an aqueous sodium chloride solution.

Sodium β-alumina is an example of a mechanically hard material that is good ionic conductor. In this case the rigid and dense Al_2O_3 slabs are bridged by a sparse array of O^{2-} ions. The plane containing these bridging ions also contains Na^+ ions which can move from site to site because there are no major bottlenecks to hinder their motion. Many similar rigid materials having planes or channels through which ions can move are known, and are called framework electrolytes.

Another closely related material, sodium β"-alumina, has even less restricted motion of ions than in β-alumina and it has been found possible to substitute dipositive cations such as Mg^{2+} or Ni^{2+} for Na^+. Even the large lanthanide cation Eu^{2+} can be introduced into β"–alumina, although the diffusion of such ions is slower than their smaller counterparts.

TYPES OF EDTA TITRATIONS

The most important procedures employed for the titration of metal ions with EDTA are :

1. **Direct Titrations :** The solution containing the metal ion to be determined is buffered to the desired pH and titrated directly with the standard EDTA solution. In order to avoid the precipitation of the hydroxide of the metal, complexing agent such as tartrate or citrate is added. At the equivalence point the magnitude of the concentration of the metal ion being determined decreases sharply.

It is accomplished by the change of colour of a metal indicator which responds to changes in pn. The point may also be determined by physical methods such as amperometric, conductometric, spectrophotometric, etc.

2. **Back Titrations:** A direct titration may not always applicable due to

   (a) It may not be possible to keep the metal in solution because of too high a pH or the presence of a precipitating reagent.

   (b) The rate of formation of the EDTA complex of the metal to be titrated may be too slow.

   (c) There may be no indicator available which gives a suitable colour reaction with the metal ion and an instrumental end point detection may not be feasible.

   (d) The metal ion may not form a sufficiently stable complex with the indicator (or none at all) to enable the titration to be made.

   Under these critical conditions the use of back titration is made which solves the problem. A measured amount of EDTA is added in excess of that necessary for reaction with the metal to be titrated. The titration conditions are then adjusted, an indicator is added and the excess of EDTA is back titrated with a suitable salt often used for this purpose.

3. **Displacement or Substitution Titrations:** Such titrations may be employed for metal ions that do not react (or react unsatisfactorily) with a metal indicator, or for metal ions which form EDTA complexes that are more stable than those of other metals such as magnesium and calcium.

   In such cases metal cation M^{n+} to be determined may be titrated with magnesium or zinc complex of EDTA, when the following reaction occurs :

$$M^{n+} + Mgy^{2-} \rightleftharpoons (My)^{(n-4)} + Mg^{2+}$$

   The amount of Mg^{2+} or Zn^{2+} ions which becomes free is equivalent to the cation present and can be titrated with a standard solution of EDTA and a soluble metal indicator.

An interesting application is the titration of calcium.

4. **Alkalimetric Titration:** When a solution of disodium ethylenediamine tetraacetate, Na_2H_2y, is added to a solution containing metallic ions, complexes are formed with liberation of two equivalents of hydrogen ions.

$$M^{+n} + H_2y^{2-} \rightleftharpoons My^{(n-4)} + 2H^+$$

The H^+ thus set free can be titrated with a standard solution of a base using an acid-base indicator or potentiometric end point. Alternatively an iodate-iodide mixture is added as well as the EDTA solution and the liberated iodine is titrated with a standard sodium thiosulphate solution.

The solution of the metal to be evaluated must he accurately neutralised before titration. This is often a difficult matter on account of the hydrolysis of many salts which forms the weak feature of alkalimetric titration.

5. **Miscellaneous Method:** Exchange reactions between the nickelo cyanide ion $[Ni(CN)_4]^{2-}$ and the element to be determined whereby nickel ions are set free, have a limited applicability. Silver, palladium, and gold which themselves cannot be titrated complexometrically, can be evaluated in this.

$$[Ni(CN)_4]^{2-} + 2Ag^+ \rightleftharpoons 2[Ag(CN)4]^- + Ni^{2+}$$

$$[Ni(CN)_4]^{2-} + Pd^{2+} \rightleftharpoons 2[Pd(CN)_4]^{2-} + Ni^{2+}$$

COMPLEXONES (LARGELY) EDTA

Complexones are a group of amino polycarboxylic acids possessing the characteristic group

$$-N\begin{matrix} \diagup CH_2COOH \\ \diagdown CH_2COOH \end{matrix}$$

and they were so named by Schwarzenbàch in 1945. The simplest possible complexone is iminodiacetic acid $HN.(COOH)_2$. The other complexones in general may be regarded as derivatives of this acid. Among the important of these studies are:

1. Methyl imino acetic acid $CH_3N\begin{matrix} \diagup CH_2COOH \\ \diagdown CH_2COOH \end{matrix}$

2. Phenyliminoacetic acid 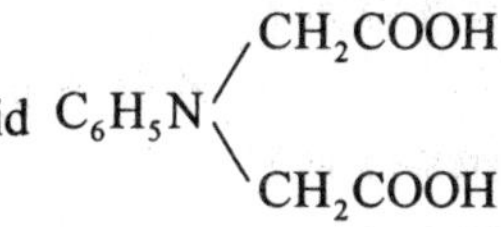

3. Nitrilo triacetic acid (NTA) 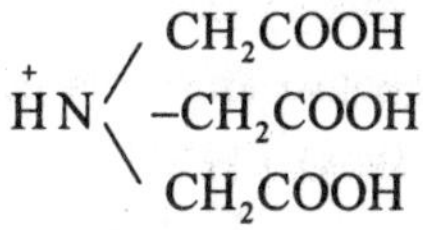

4. Ethylene diaminetera acetic acid

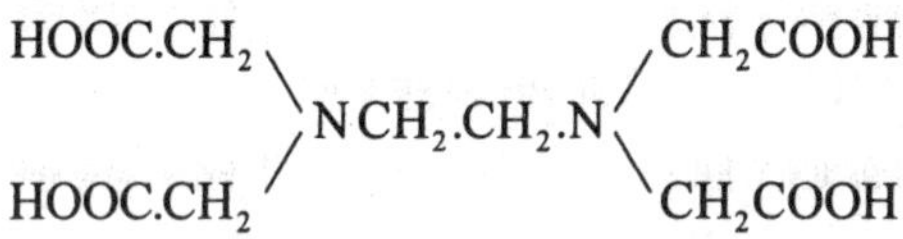

Today atleast 20 compounds of the complexones are known and their properties have been studied principally by Schwarzenbach and coworkers. Such compounds possess remarkable analytical property by virtue of their ability to form complex compounds with most of the cations. The compounds are of chelate type, *i.e.,* the cations become bound in ring structures.

Among the complexones reported in literature, ethylenediaminetera-acetic acid (EDTA) occupies a pre-eminent position. It is undoubtedly one of the most versatile complexones reported in the literature.

Ethylenediamine tetraacetic acid has the following structure:

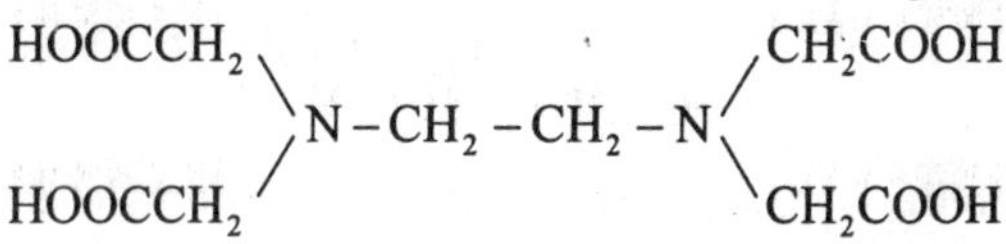

It is a weak acid for which the values of pK are respectively $pK_1 = 2.0$, $pK_2 = 2.67$, $pK_3 = 6.16$ and $pK_4 = 10.26$ at 20°C. Here K is the overall dissociation constant of EDTA and K_1, K_2, K_3 and K_4 represent the dissociation constants when EDTA gives one, two, three or four H^+ by dissociation, pK_1, pK_2, pK_3 and pK_4 are the logarithmic or exponential methods of expressing small quantities of K_1, K_2, K_3 and K_4.

$$pK_1 = -\log K_1 = \log \frac{1}{K_1}$$

On the basis of these values of dissociation constants it has been shown that in EDTA, two hydrogen atoms are held in the form of zwitter ions. Therefore, the following structure is now assigned to EDTA.

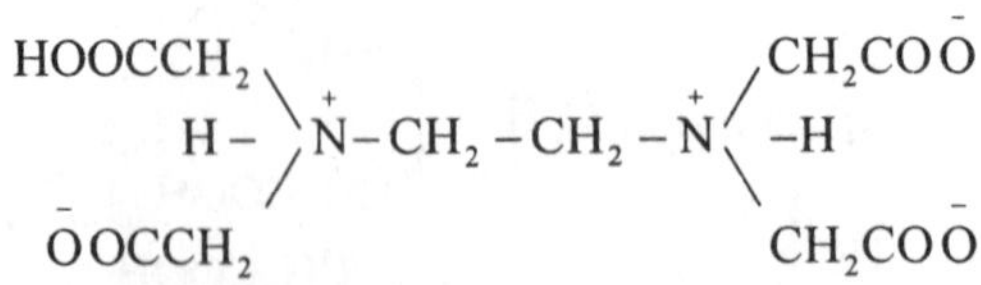

From the structure of EDTA it is clear that in its molecule, in addition to four acidic hydrogen, each nitrogen atom has an unshared pair of electrons. Therefore, a molecule of EDTA has six potential sites for bonding with a metal ion.

REACTIONS AND PROPERTIES OF EDTA

These can be considered in number of headings

1. **Complexes with Metals:** Complexes are formed with nearly every metal ion carrying more than unit positive charge and even with alkali metals like sodium and lithium. .
2. **Stoichiometry of Complex Formation:** Complexes are formed stoichiometrically (1:1) and in nearly every case, instantaneously. It is, therefore, possible to carry out direct complexometric titration with the reagent.
3. **Complex Stability and pH:** pH has a marked effect on the stability of various EDTA metal complexes. The complexes of divalent metals are stable in ammoniacal solution but tend to decompose in acid medium, while those of trivalent transition metals are stable even in strong mineral acid. These complexes are only slightly ionised.
4. **Solubility of Complexes:** Nearly all the complexes are soluble in water, leaving a few, which do not dissolve even in the excess of EDTA, ammonia or dilute acid. Generally, the complexes are not extracted into immiscible organic solvents.
5. **Complex Colour:** Colourless metal ions give colourless EDTA complexes while those of coloured ions almost invariably possess deep colours.
6. **Solubility of EDTA:** It has itself limited solubility in water and can be precipitated from acid solution.
7. **Stability of EDTA:** It is destroyed by moderately powerful oxidising agents, *e.g.*, hot dichromate and ceric solutions.

8. **Masking Reaction:** EDTA is a very unselective reagent because it complexes with numerous di, tri-, and tetravalent cations. By the use of masking agents some of the cations in a mixture can often be 'masked' so that they can no longer react with EDTA. An effective masking agent is the cyanide ion.

9. **EDTA as a Primary Standard:** EDTA may be prepared sufficiently pure for use as a primary standard solution which exhibits remarkable stability. EDTA itself has only unlimited solubility in water and can be precipitated in hot solution.

10. **Formation and Structure of EDTA Metal Chelates:** EDTA is a hexadenate ligand due to the most favourable combinations of the four carboxylic group (for salt formation) and two amino nitrogen (for co-ordination) in its structure. The complexes formed are of chelate type, *i.e.,* the cations become bound in ring structures. Salt-like bonds are formed in ring structure by the replacement of H+ ion from the –COOH groups and ring closing takes place by co-ordination of lone pair of electrons by nitrogen atom.

The divalent metals react with disodium salt of EDTA according to following reaction

$$Na_2H_2y \rightleftharpoons 2Na^+ + H_2y^{2-}$$

$$M^{2+} + H_2y^{2-} \rightleftharpoons My^{2-} + 2H^+$$

USES OF EDTA IN INORGANIC ANALYSIS

1. **EDTA as Titrant:** The use of EDTA as a titrant avoids the difficulty of most complexometric titrations in that a single complex species is formed instead of the usual stepwise formation of complexes which occurs with many complex forming ions or molecules. The formation of a succession of complexes with a single cation makes difficult the location of the end point in a titration of this type and that is why EDTA is now widely used for the titrimetric determination of as many as 40 cations.

   Schwarzenbach outlined three main methods of application as a titrant :

   1. Alkalimetric method.
   2. pH effect method.

3. Metal specific indicator method.

(i) **Alkalimetric Method:** The fundamental reaction involved is

$$Na_2H_2y + M^{2+} \rightleftharpoons Na_2My + 2H^+$$

The hydrogen ions liberated measures the concentration of the metal which has been complexed.

(ii) **pH Effect Method:** Here the tetrasodium salt of EDTA which is strongly alkaline is used as titrant.

$$Na_4y + M^{2+} \rightleftharpoons Na_2My + 2Na^+$$

$$Na_4y + 2H^+ \rightleftharpoons Na_2H_2y + 2Na^+$$

(iii) **Metal Specific Indicator Method:** Mostly EDTA titrations are performed with indicators. These are chelating agents whose metal complexes have different colour from the reagent. Such indicators are commonly called "Metalochromic indicators". The requirements of a metal indicator are

- (i) The metal-indicator must have a different colour from metal indicator compound.
- (ii) The metal-indicator complex must be formed under the same conditions as EDTA-metal complex.
- (iii) The indicator must be very sensitive to metal ions so that colour change occurs as near to the equivalence point as possible.
- (iv) The metal indicator complex must be rather less stable than the EDTA complex.

The indicators most commonly used are the following

(i) Eriochrome black T

(ii) Murexide

(iii) Pyrocatechol violet

(iv) Dithiozone

(v) l-(2 pyridyl azo)-2-naphthol (PAN).

2. **Determination of Metals:** Method for the determination of metals was first developed by Schwarzenbach. Metals like aluminium, iron, copper, cadmium, zinc, cerium, chromium, mercury, bismuth,

lead, uranium, scandium, rare earths, etc., have been determined directly or indicrectly.

3. **Determination of Anions:** In addition to the determination of metal ions, EDTA may also be used for the determination of certain anions. The anions which have been determined are arsenate, chromate, ferrocyanide, fluoride, pyrophosphate, sulphate, etc.
4. **Determination of Water Hardness:** A particular application of EDTA which has received wide attention in its use is the determination of hardness in water on account of the ease with which calcium and magnesium can be titrated. Considerable advantages are obtained over older methods when EDTA is used.
5. **Use of EDTA in the Analysis of Technological Materials:** EDTA has been used extensively in the analysis of commercial materials such as alloys, cements, coins, leather, mineral oils, pertrol, slags, many ores and rocks.
6. **EDTA as a Masking Agent or Sequestrating Reagent:** EDTA may be used to prevent interference of certain metals in the determination of other ions with conventional reagents. For example, uranium can be separated from numerous other ions by precipitation with oxine from an EDTA solution at pH 5.3. Copper has been masked with EDTA in the polarographic determination of antimony in alloys.

The term sequestration is used in a restricted sense to describe the suppression of the precipitate reactions of calcium and magnesium. For example, the precipitation of soaps as their calcium and magnesium salts by hard water can be checked by transforming these metal into ions of soluble complexes with EDTA. Again, the formation of scale in the boilers can be prevented by complex formation with EDTA.

MISCELLANEOUS APPLICATIONS OF EDTA

A number of miscellaneous applications of EDTA has been also reported. Some workers have described the EDTA titrations of calcium, cobalt, lead, zinc and manganese in paint driers in alcoholic medium. The gravimetric determination of magnesium using EDTA as a precipitant has also been described by Bvicker (1957).

Beside the above uses of EDTA, its use as a chromogenic reagent in spectrophotometric procedures, its role in polarography and ion exchange technique make EDTA an important reagent.

Importance of EDTA Complexometric Titrations

1. Coprecipitation errors associated with precipitation titrations do not exist in these titrations.
2. Many metals can be determined with a single chelating agent.
3. Selectivity among the metals being achieved by control of pH at which titrations are per formed.

ADSORPTION INDICATORS

In 1924, Fajans utilised useful type of indicators for certain precipitation titrations, known as adsorption indicator.

Surface of certain indicators possess a property of adsorption. At the equivalence point, certain ions are adsorbed and a typical orientation is developed at the surface which imparts a characteristic colour change at the end point. Due to this phenomenon of adsorption these are termed as adsorption indicators. Hence *"Adsorption indicator is such a substance on whose surface adsorption takes place at equivalence point, resulting in a colour change to detect the end point"*.

The substances used as adsorption indicators are :

(i) **Acid dyestuffs of the fluorescein series :** Examples of this series are fluorescein, eosin, dichlorofluorescein. These are utilised as sodium salts.

(ii) **Basic dyes :** Examples of rhodamine series are generally used as halogen salts.

Example : Fluorescein is used as indicator during the titration of chloride with $AgNO_3$.

When a chloride solution is titrated with a solution of silver nitrate, the precipitated silver chloride adsorbs chloride ions (primary adsorbed layer). These negatively charged chloride ions adsorb positively charged ions present in sodium chloride solution (secondary adsorbed layer). For this see Fig. 69.1(A).

$$AgNO_3 + KCl \rightarrow AgCl \downarrow + KNO_3$$

(a) AgCl precipitated in the presence of excess of Cl^-.

(b) AgCl precipitated in the presence of excess of Ag^+.

If fluorescein is also present in the solution, the negative fluorescein ions, which are much more strongly adsorbed than the nitrate, are immediately adsorbed. This gives red colour due to the formation of complex of silver and fluoresceinate ion on the surface. Thus, at the end point red colour is obtained.

Another view is that during the adsorption of the fluorescein ion, a rearrangement of the structure of the fluorescein occurs with the formation of a coloured substance. This colour change takes place only at the surface of the precipitate.

Conditions for the choice of a suitable adsorption indicator :

(i) The indicator ion must be of opposite charge to the ion of precipitating agent.

(ii) The indicator ion should be adsorbed before the particular compound has been completely precipitated but it should be strongly adsorbed immediately after the equivalence point.

(iii) The precipitate should separate as far as possible in the colloidal condition.

The theory and the application of such indicators have been reviewed by *Mehrotra et al.* (1964).

Extractive Indicators

The new technique of employing an extractive indicator in complexometric titrations was used for the first time by Guill and others (1951) in the titration of zinc against potassium ferrocyanide. *An extractive indicator is one whose solution is used in an organic solvent which is immiscible with the aqueous phase and hence forms a separate layer.* The volume of non-aqueous layer being much less than aqueous phase and hence due to the increase in concentration effect, the end point is more efficiently marked.

Although, the extractive method of end point detection introduces an additional step in an analysis, this disadvantage is overweighed by the following noted advantages:

1. The end point is no way affected by the presence of an reactive coloured ion or highly coloured metal- EDTA complexes present in the aqueous phase.

2. The sensitivity of the end point is increased by the use of relatively small volume of extracting solvent.

Extractive indicators are employed either as acid-base or metal indicator. In the case of acid-base indicators, the phase transfer equilibria are reached very slowly and, therefore, they offer little advantage over others which function satisfactorily in homogeneous media. Extractive indicators are, however, of great value as metal indicators in complexometric titrations. Among extractive indicators reported dithizone is most prominent. Very recently diphenyl thiovioluric acid has been reported as a new extractive indicator by Mehral in 1972.

Screened Indicators

Some times single indicators do not give a sharp colour change during acidimetric and alkalimetric titrations. It is observed that addition of an indifferent dyestuff often improves the sharpness of the colour change. The dyestuff does not undergo change itself with change of pH but is present to screen out the transmitted light. Consider the methyl orange indicator which is screened with xylenecyanol FF. The absorption of alkaline methyl orange is almost blue and for acid methyl orange blue green, so that the observed colour orange is from yellow to orange red.

Xylenecyanol FF has an absorption in the yellow orange region. A mixture of the two in alkaline medium gives a green coloured solution while the acidic yields a red to magenta colour. The colour change of green to magenta through a neutral grey colour is easier to detect than yellow to red through orange.

ARRHENIUS CONCEPT—THE WATER IONS SYSTEM

According to Arrhenius theory of ionisation, all substances which give hydrogen ions when dissolved in water could be called acids while, those which ionise in water to form hydroxyl ions, coula be called bases.

Examples of Arrhenius acids are HCl, H_2SO_4, CH_3COOH, etc. Examples of Arrhenius bases are NaOH, NH_4OH, etc.

Acid-base equilibria could now be treated in a quantitative way and acid-base neutralisation could he considered to result from the union of the hydrogen ions with hydroxyl ions to form water molecules,

$$H^+ + OH^- \longrightarrow H_2O$$

When we write a general equation.

$$HA \longrightarrow H^+ + A^-$$

for an acid, suggesting the formation of a group, H^+, but this is not very correct as free protons could exist only when gaseous hydrogen atoms are ionised. In the presence of protophilic liquids such as water, unattached proton could not exist; in water a proton gets attached to a number of water molecules by co-ordinate linkages, releasing a very large amount of energy. The proton in its hydrated state is usually expressed by H_3O^+, even though the number of water molecules attached to one proton is more than one. The hydrated proton is called the hydroxonium or hydronium ion. On the other hand hydroxyl ions are also hydrated in water through hydrogen bonds and are represented by OH- formula only because of convenience.

Successes

(1) Arrhenius concept has been invaluable in elucidating the behaviour of aqueous solutions. The constant heat of neutralisation of a strong acid by a strong base is readily explained in the light of this concept.

(ii) This theory correlated the catalytic action of acids with the concentration of hydrogen ions.

(iii) This concept also lends to quantitative determinations of acid or base strengths front the determination of following equilibrium relation:

$$K = \frac{{}^{a}H^{+}\,{}^{a}H^{-}}{{}^{a}HB}$$

Limitations

(i) The Arrhenius concept lacks in generality.

(ii) The most serious limitation is that it deals with dissociation and acid-base reactions in aqueous medium only and does not mention the role of non-aqueous solvents in the dissociation of acids and bases into ions. According to this concept, HCl is regarded as an acid only when dissolved in H_2O and not in some other solvents like benzene, chloroform, etc. or when it exists in the gaseous form.

(iii) This concept restricts bases merely to hydroxides. Thus, even metal oxides would not be regarded as bases.

(iv) According to this concept, a neutralisation process is restricted to aqueous medium alone but many non-aqueous solvents like liquid NH_3 liquid SO_2, etc. are known to undergo comparable acid-base neutralisation reactions.

(v) It cannot explain the acidic behaviour of certain salts such as $AlCl_3$ in water.

(vi) This concept does not elaborate the fate of dissociated hydrogen' ions in solution because of their high hydrational energy (1075 kj/mole).

(vii) A proton (H^+) is non-existent and is generally present as the hydrated H_3O^+ ion or H_9O_{4+} species.

(viii) A proton is also formed in other solvents like NH_3, ROH, etc.

(ix) There are nonprotonic and nonhydroxylic compounds which may yield H^+ or OH ions in water.

$$SO_3 + H_2O \rightarrow H_2SO_4 \rightleftarrows H^+ + HSO_4^-$$

$$CaO + H_2O \rightarrow Ca(OH)_2 \rightleftarrows Ca^{2+} + 2OH^-$$

THE PROTON-DONORACCEPTORSYSTEM

Bronsted and Lowry in 1923 proposed a new concept to acids and bases which is independent of solvents. According to atom.

"An acid is any hydrogen contained substance (a molecule or a cation or an anion) whereas a base is any substance (a molecule or cation or an anion) that can accept a proton from any other substance."

From the above definition, it follows that an acid is a proton donor and base is a proton acceptor. It also follows from this definition that it is not essential for bases to have OH^- groups as demanded by the Arrhenius theory.

A neutralisation process, therefore, involves a release of a proton by the acid, and the acceptance of that proton by a base. This concept makes acid-base reactions also to be known as protolytic reactions independent of solvent.

$$\underset{\text{acid}}{H} : Cl + \underset{\text{base}}{NH_3} \rightarrow Cl^- + NH_4^+$$

$$\underset{\text{acid}}{NH_4^+} + \underset{\text{base}}{OH^-} \rightarrow NH_3 + H_2O$$

Bronsted and Lowry acids and bases are of the following types :

Bronsted-Lowry Acids

$$\begin{bmatrix} \text{Molecular : HCl} \rightarrow H^+ + Cl \\ \text{Cationic : } [A\,(H_2O)_6]^{3+} \rightarrow H^+ + [A\,(H_2O)_5OH]^{2+} \\ \text{Anionic : } H_2PO_4^- \rightarrow H^+ + HPO_4^{2-} \end{bmatrix}$$

Bronsted-Lowry Bases

$$\begin{bmatrix} \text{Molecular : Py} + H^+ \rightarrow\,]PyH]^+; \text{ where Py = Pyridine} \\ \text{Cationic : } [A(H_2O)_5OH]^{2+} + H^+ \rightarrow [A(H_2O)_6]^{3+} \\ \text{Anionic : } PO_4^{3-} + H^+ \rightarrow HPO_4^{2-} \end{bmatrix}$$

Substances like H_2O and HCO_3 ions who donate as well as accept protons.

$$H_2O \rightleftarrows H^+ + OH^- \text{ (donation)}$$

$$H_2O + H^+ \rightleftarrows H_3O^+ \text{ (acceptance)}$$

$$HCO_3^- \rightleftarrows H^+ + CO3 \text{ (donation)}$$

$$HCO_3^- + H^+ \rightleftarrows H_2CO_3 \text{ (acceptance)}$$

are classed as amphoteric substances

The Bronsted-Lowry concept is quite useful For example:

(i) This concept also takes into account protonic solvents like NH_3, H_2SO_4, etc.

$$NH_3 + NH_3 \rightleftarrows NH_4^+ + NH_2^-$$

$$H_2SO_4 + H_2SO_4 \rightleftarrows H_3SO_4 + HSO_4$$

(ii) All-proton transfer reactions may be treated as follows:

$$NH_4^+ + S^{2-} \rightarrow NH_3 + HS$$

Conjugate Acids and Bases

A study of the following reactions for processes occurring in water brings out another important feature of this concept about acids and bases:

$$CH_3COOH + OH^- \rightleftarrows HOH + CH_3COO^-$$

$$H_2O + H_2O \rightleftarrows H_3O^+ + OH^-$$

$$NH_4^+ + CO_3^{2-} \rightleftarrows HCO_3^- + NH_3$$

$$H_2O + CO_3^{2-} \rightleftarrows HCO_3^- + OH^-$$

From the above reactions, it is seen that when a molecule or an ion loses a proton, it is necessary to convert into a form which would accept a proton, to reform the original acid. The latter form is therefore, a base and it is called a linked or conjugate base to the initial acid. Such pairs of substances which are formed from one another by the gain or loss of a proton are called conjugate acid-base pairs.

Acetic acid and acetate ion, H_2O and OH^- ion and NH_4^+ ion and NH_3, H_3O^+ ion and H_2O, HCO_4 ion and CO_3^{2-} ion are all examples of conjugate pairs of acids and bases. Just as for every weak acid we have a conjugate base, for every weak base we have a conjugate acid which is formed by the addition of a proton to the base. The strong acids and bases are not covered by this statement for the lack of the corresponding proton combination.

Furthermore, a substance like HNO_3 behaving as an acid or a base depends on the solvent uses. In liquid NH_3, H_2O and even dry CH_3COOH, HNO_3 acts like an acid by donating protons, but in dry HF medium, it accepts protons, and behaves like a base.

Relative Strength of Acids and Bases

According to Bronsted concept, the strength of an acid is regarded as a measure of its tendency to donate proton. Thus, a strong acid is that one which has a strong tendency to donate a proton. Similarly, a strong bases is that one which has a strong tendency to accept a proton.

Generally, two methods are employed to compare the relative strength of given acids. These are as follows

(i) In the first method we compare the dissociation constants or ionisation constants of acids/bases to get idea about their strength. Higher the dissociation constant, greater the strength of acid/base. For example, we are interested in comparing the acidic strength of CH_3COOH and HCN. The values of dissociation constant of CH_3COOH and HCN are 1.8×10^{-5} and 4.0×10^{-10} respectively.

$$CH_3COOH + H_2O \rightleftarrows H_3O+ + CH_3COO^-(K_a = 1.8 \times 10^{-5})$$

$$HCN + H_2O \rightleftarrows H_3O^+ + CN^- \quad (Ka = 4.0 \times 13^{-10})$$

As the value of K_a of CH_3COOH is greater than that of HCN, it means that CH_3COOH is a stronger acid than HCN and CN^- ion is a stronger base than CH_3COO^- ion

(ii) The second method is known as competitive protolysis method. In this method, one acid is added to the conjugate base of another acid, and, then, one has to determine the equilibrium concentrations experimentally.

For example, if $NaOC_2H_5$ is added to H_2O, it is exprimentally found that $C_2H_5O^-$ has good tendency to react completely with the H_2O to from C_2H_5OH and OH^- ion. It means that $C_2H_5O^-$ is a stronger base than OH^- and H_2O is a stronger acid than C_2H_5OH.

$$\underset{\text{Stronger acid}}{H_2O} + \underset{\text{Stronger base}}{C_2H_5O^-} \rightleftarrows \underset{\text{Weaker acid}}{C_2H_5OH} + \underset{\text{Weaker base}}{OH^-}$$

It is important to mention here that acidity increases with the increase of positive charge whereas the basicity increases with the increase of positive charge. Thus, $[Fe(H_2O)_6]^{3+}$ is a stronger acid than $[Fe(H_2O)_6]^{2+}$ ion and $[Ni(OH)_4]^{2-}$ is a stronger base than $[Ni(OH)_4]^-$ ion.

Successes

(i) This concept shows that acid base phenomena are dependent on the solvent.

(ii) The concept covers a large number and variety of substances under base, for example, hydroxide ion, amide ion, ethoxide ion, pyridine, alcohols, acetate ion, bisulphate ion and many others.

(iii) Bronsted-Lowry concept is useful in accounting for the hydrolysis of salt solution. A solution of $FeCl_3$ is acidic because the proton donor ability of hydrated ferric ion exceeds the proton acceptor ability of the chloride ion and measurable excess of $[Fe(H_2O)_x]^{3+}$, ion appears in the solutions.

$$FeC1_3 + {}_xH_2O \rightleftarrows [Fe(H_2O)_x]^{3+} + 3Cl^-$$

Limitations

This theory serves excellently in protonic solvents but fails in case of some obvious acid-base reaction. For example, the formation of

H_2SO_4 by the reaction of SO_3 and H_2O, the acidic participant SO_3 in this case has no protons to give.

LEVELLING SOLVENTS

The apparent strength of a protonic acid has been found to depend upon the solvent in which the acid is dissolved. When all the acids which are stronger than H_3O^+ (*e.g.*, $HClO_4$, H_2SO_4, HCl and HNO_3) are added to H_2O, they lose a proton to H_2O forming H_3O^+ ion. All these acids have equal strength because all these acids have been levelled to the strength of H_3O^+ ion which remains in solution and is common to all such solutions. This phenomenon in which the strength of all the acids becomes equal to that of H_3O^+ ion is known as levelling effect of the solvent *i.e.*, water. Here water is known as a levelling solvent for all these acids.

Solvent	*Acid*	+ *Base*	→ *Salt*	+ *Solvent*
H_2O	H_3O^+ (HBr)	OH^- (KOH)	K^+, Br^-	$2H_2O$
C_2H_5OH	$C_2H_5OH_{2+}$ (HBr)	$O^-C_2H_5$ (KOC_2H_5)	K^+, Br^-	$2C_2H_5OH$

The solvents in which complete transfer takes place are known as levelling solvents. It means that the solvents in which the solute is 100% ionised are known as levelling solvents. For example, liquid NH3 acts as a levelling solvent for HF and HCI because both these are -100% ionised in liquid NH_3 to give-100% NH_4^+ ions.

In H_2O, HF undergoes partial ionisation while HCl and HBr undergo –100% ionisation. Thus, H_2O is a differentiating solvent for HF but for HCl and HBr it is a levelling solvent.

Other solvents may have their own levelling effects, too. Thus, benzoic acid is a very weak acid in aqueous solution. But in liquid ammonia it acts as a very strong acid. In general, it may be expected that acidic solvents (like acetic acid) will exert less levelling effect on acid strengths (and similarly-basic solvents on base strengths).

Relative strengths of the common mineral acids have been obtained by conductace measurement in anhydrous acetic acid. The order found is

$$HClO_4 > HBr > H_2SO_4 > HCl > HNO_3$$

SOLVENT SYSTEM CONCEPT

Franklin extended the dissociation principles to non-aqueous solvents and realised that non-aqueous solvent molecules may also dissociate into two oppositely charged ions. According to him, if an acid releases H^+ in water to form H_3O^+ ions then an ammonium salt, NH_4Cl, by releasing NH_4^+ in liquid ammonia, must also be termed an acid.

If NaOH, by virtue of its releasing OH^- ion in aqueous medium, is termed a base in that medium, then $NaNH_2$ should also be termed a base in liquid NH_3 since it releases the amide, NH_2^- ion. The dissociation (or auto-ionisation) of non-aqueous solvent is responsible in such solvents.

$$NH_3 \rightleftarrows H^+ + NH_2^-$$

According to Franklin, an acid is a substance which by dissolution in the solvent forms the same cation as does the solvent itself due to autoionisation. A base is one that gives on dissociation in the solvent the same anion as does the solvent itself on its own ionisation.

Solvent	*Acid +*	*Base →*	*Salt +*	*Sovent*
H_2O	H_3O^+ (HBr)	OH^- (KOH)	K^+, Br^-	$2H_2O$
C_2H_5OH	$C_2H_5OH_2^+$ (HBr)	$O^-C_2H_5$ (KOC_2H_5)	K^+, Br^-	$2C_2H_5OH$

The solvent system concept to non-protonic solvent *e.g.*, SO_2, $COC1_2$ in case of acids and bases was applied by Cady and Elsey.

According to them, substances which increase the concentration of the cation characteristics of the solvent are acids, whereas substances which increase the concentration of the anion characteristics of the solvents are bases.

In N_2O_4 as a solvent, substances such as NOCl which yield NO^+ ions behave as acids, and substances such as $NaNO_3$ which yield NO_3^- ions behave as bases.

The auto-ionisation of some non-protonic solvents is given below:

Acid Base *Acid Base*

$$SO_2 + SO_2 \rightleftarrows SO^+ + SO_3$$

$$COCl_2 + COCl_2 \rightleftarrows COClCOC_2^+ + Cl^-$$
$$N_2O_4 + N_2O_4 \rightleftarrows 2NO^- + 2NO_3^+$$
$$BrF_3 + BrF_3 \rightleftarrows BrF_2 + BrF_4^-$$

Just as with Arrhenius concept neutralisation is reaction between an acid and a base to produce a salt and the solvent. Neutralisation reactions in non-aqueous media are given below:

Acid *Base* *Salt* *Solvent*

In liquid NH_3: $NH_4Cl + NaNH_2 \rightleftarrows NaCl + 2NH_3$

In liquid SO_2: $SOCl_2 + Na_2SO_3 \rightleftarrows 2NaCI + 2SO_2$

It is clear from the following reactions that there is complete analogy between solvolytic and amphoteric behaviours in aqueous and non-aqueous solvents.

Solvolytic Behaviour

$$\left[\begin{array}{l} \text{In liq. } NH_3 : AlCl_3 + NH_3 \rightarrow [Al(NH_2)]^{2+} + H^+ + 3Cl^- \\ \text{In } H_2O \quad : AlCl_3 + H_2O \rightarrow [Al(OH)^{2+} + H^+ + 3Cl^- \end{array}\right.$$

Amphoteric behaviour:

$$\left[\begin{array}{l} \text{In liq } NH_3 : Zn(NH_2)_2 + 2NH_2^- \rightarrow [Zn(NH_2)_4]^{2-} \\ \quad \text{In } H_2O : Zn(OH)_2 + 2OH^- \rightarrow [Zn(OH)_4]^2 \end{array}\right.$$

Success

This general theory of solvent system concept possesses the obvious advantage of relating acid-base behaviour in aqueous and non-aqueous solvents, both protonic and nonprotonic.

Limitations

(i) It limits acid-base phenomena to solvent systems only.

(ii) The concept does not explain the neutralisation occurring without the absence of solvent.

(iii) It does not explain the acid-base processes which may occur in the absence of solvent.

(iv) In low dielectric constant solvents, the existence of ions is not attainable.

(v) Emphasis has been laid mainly on the chemical properties of the solvent. However, the physical properties have been completely ignored.

(vi) Too much stress has been laid mainly on the ionic reaction which actually even dielectric affects would be able to explain the reactions easily. For instance, Gutman and his coworkers (1952-60) have explained reactions between the chloride ion acceptors ($FeCl_3$,$SnCl_4$, etc.) and chloride ion donors (Me_4NCl) in $OPCl_3$ as follows:

$$OPCl_3 \rightleftarrows OPCl_2^+ + Cl^- \text{ (autoionisation)}$$

$$Me_4NCl \rightleftarrows Me_4N^+ + Cl^- \text{ (base)}$$

$$OPCl_3 + FeCl_3 \rightleftarrows OPCl_2^+ + FeCl_4^- \text{ (acid)}$$

However, Meek and Drago (1962) showed that the reaction can be carried out equally well in triethyl phosphate medium without involving chloride ions.

It means that the dielectric constant of the medium is more important than the ionisation reaction of the solvent.

Thus the solvent system concept for acid and bases simply extends, but does not necessarily improve, the Arrthenius waterions system.

LUX-FLOOD CONCEPT

Lux proposed a theory to explain the acid-bases concept of oxide system. According to the concept, a base is any material which gives up oxide ions and an acid is any material which gains oxide ions. Some typical reactions are

$$\text{Base} \rightleftarrows \text{Acid} + {}_xO^{2-}$$

$$SO_4^{2-} \rightleftarrows SO_3 + O^{2-}$$

$$CaO \rightleftarrows Ca^{2+} + O^{2-}$$

Hence, according to the concept, a bases is an oxide donor and acid is an oxide ion acceptor. This view is particularly useful to high temperature chemistry, as in the fields of ceramics and metallurgy.

For example, the following reactions

$$CaO + SiO_2 \rightarrow CaSiO_3$$

$$PbO + SO_3 \rightarrow PbSO_4$$

involving basic oxides (CaO,PbO) and acidic oxides (SO_3, SiO_2) are reacting to form salts. Furthermore, according to this theory amphoteric substances are those which show both a tendency to take up or give up oxide ions depending upon the circumstances, *i.e.*,

$$Na_2O + \underset{\text{Acid}}{ZnO} \rightarrow 2Na^+ + ZnO^2$$

The oxide transfer picture due to Lux can be extended to any negative ion, *i.e.*, halide, sulphide, etc.

$$3NaF + AlF_3 \xrightarrow[\text{temperature}]{\text{high}} 3Na^+ + AlF_6^{2-}$$

The Lux-flood definition can be extended to include transfer of any anion *e.g.*, halide, sulphide, etc.

Limitations

The limitations of such a concept are obvious. This view can be reduced to one part of the more general theories of Lewis and Usanovich.

LADY-ELSEY CONCEPT

Bases on the characteristic cation or anion of the solvent which is produced by it with the solute, a more general definition of acids and bases applicable to all solvents was provided by Cady and Elsey. Consider, for example.

(i) $$H_2O + H_2O \rightarrow H_3O^+ + OH^-$$

(ii) $$NH_3 + NH_3 \rightarrow NH_4^+ + NH_2^-$$

(iii) $$HF + HF \rightarrow H_2F^+ + F^-$$

(iv) $$N_2O_4 + N_2O_4 \rightarrow 2NO^+ + 2NO_3^-$$

(v) $$CH_3COOH + CH_3COOH \rightarrow CH_2COOH_2^+ + CH_3COO^-$$

The characteristic cations furnished by water, ammonia, hydrogen fluoride and nitrogen tetroxide from (i), (ii), (iii) and (iv) above are respectively H_3O^+, NH_4^+, H_2F^+ and NO^+. The respective characteristic anions of these solvents are OH^-, NH_2^-, F^- and NO_3^-.

According to Cady-Elsey Concept

An acid is a substance which in dissolution in solvent gives the cations characteristic of the solvent and a base is a substance which yields anion characteristic of the solvent either by direct dissolution or by reaction with the solvent.

In the light of this concept, it is seen from (i), (ii), (iii) and (iv) above that:

All substances which furnish H_3O^+ ions in water solvent, NH_4^+ ions in ammonia solvent, H_2F^+ in hydrogen fluoride solvent, NO^+ in nitrogen tetroxide and acetic acid medium are termed as bases. For example,

$$NH_4X \rightleftarrows NH_4 + X^-$$

In ammonia is an acid and $MNH_2 \rightarrow M^+ \; NH_2^-$ in it is ao base. BF_3 in hydrogen fluoride reacts as $BF_3 + 2HF \rightleftarrows H_2F^+ + BF_4^-$ forming H_2F^+ ions. It is therefore an acid.

ELECTRON-DONOR-ACCEPTOR SYSTEM OR THE LEWIS CONCEPT

Gilbert N. Lewis proposed a broader concept of acids and bases that liberated acid-base phenomena from the proton; although Lewis first proposed his system in 1923, he did little to develop it until 1938, Lewis defined a base as a substance that has an unshared electron pair with which it can form a covalent bond with an atom, molecule, or ion.

An acid is a substance that can form a covalent bond by accepting an electron pair from a base. An acid must have a vacant orbital into which an electron pair donated by a base can be accommodated. An example of a Lewis acid-base reaction that is not treated as such by any other acid-base concept

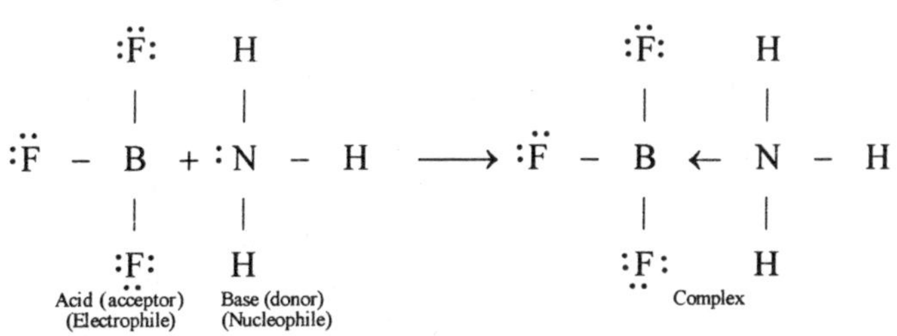

Fig. 1

Lewis acid is often described as acceptor or electrophile, and the Lewis base is called a donor and nucleophile. On the basis of this concept, cations are acid-base complexes.

Classification of Lewis acids and bases : Based on the Lewis system are essentially the same as in Bronsted-Lowry system, since any molecule or ion which accepts protons does so because of the presence of an unshared pair of electrons. For example, OH^-, CF^-. NH_3, H^3O^+, CN^- etc., are bases on the Lewis as well on Bronsted-Lowry system.

The properties of an acid might be due to the availability of an empty orbital for the acceptance of a pair of electrons. Hence Lewis definition does not attribute acidity to any particular element but rather, to a unique electronic arrangement. According to this concept, acids are classified as

1. **Molecules possessing a central atom with less than an octet of electrons :** Lewis acids of this class are the electron deficient molecules such as B in BF_3, Al in $AlCl_3$.

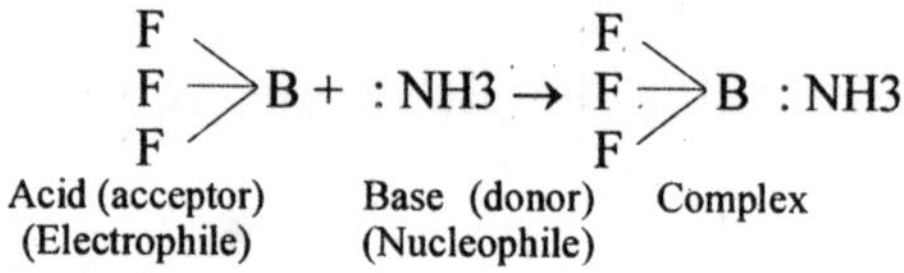

Fig. 2

2. **Molecules in which the central atom has available 'd' orbitals and may acquire more than an octet of valence electrons:** Silicon tetrafluoride and tin (IV) chloride are typical examples:

$$\underset{\text{Lewis acid}}{SiF_4} + \underset{\text{Lewis base}}{2:\ddot{\underset{..}{F}}^-:} \rightarrow \underset{\text{Complex}}{SiF_6^{2-}}\ ;\ \underset{\text{Lewis acid}}{SnCl_4} + \underset{\text{Lewis base}}{2:\ddot{\underset{..}{Cl}}^-}; \rightarrow \underset{\text{Complex}}{SCl_6^{2-}}$$

Some other examples of this type are

PF_3, PF_5, $TeCl_4$, SF_4, SeF_4, TiX_4, GeX_4,

These halides tend to form adducts with a number of organic bases.

3. **Cations :** This class of Lewis acids is made up of positively charged heavy metal ions with incomplete stable orbitals. Theoretically, all cations are potential Lewis acids because they are electron deficient. However, this property is negligible for the alkalimetal cations such as Na^+, and K^+, and is weak for the alkaline earth cations such as Ca^{2+}, Sr^{2+} and Ba^{2+}. Of the alkaline earth group, only Mg^{2+} and Be^{2+} show an appreciable tendency to behave as Lewis acid. The reaction of M^{3+} type of ion with water molecules is an example of Lewis acid-base interaction.

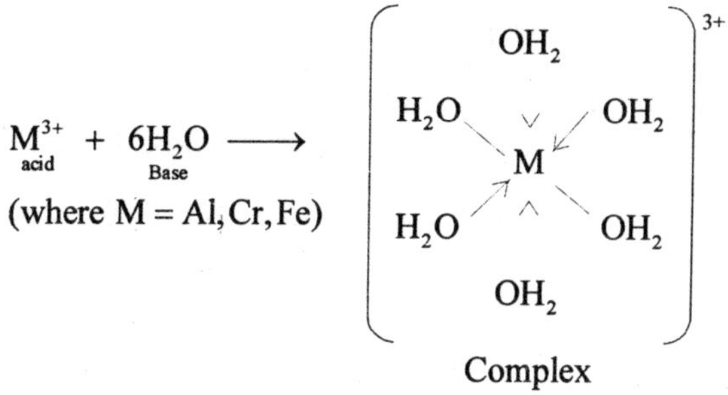

Fig. 3

The reaction of Ag^+ with NH_3 is also typical

$$\underset{\text{Lewis acid}}{Ag^+} + 2:\underset{\text{Lewis base}}{NH_3} \rightarrow \underset{\text{Complex}}{[H_3N:\ Ag\ :NH_3]^+}$$

Usually, the number of reacting nucleophiles is twice the charge on the cation.

4. **Molecules with a multiple bond between atoms of dissimilar electronegativities :** An example of such a molecule is carbon dioxide, O = C = O. The oxygen atoms are more electronegative than the carbon atom. As a result, the electron density due to the it electrons is displaced away from the carbon atom, towards the oxygen atoms. The carbon atom is electron deficient and is apt to from a bond with a Lewis base such as OH^-

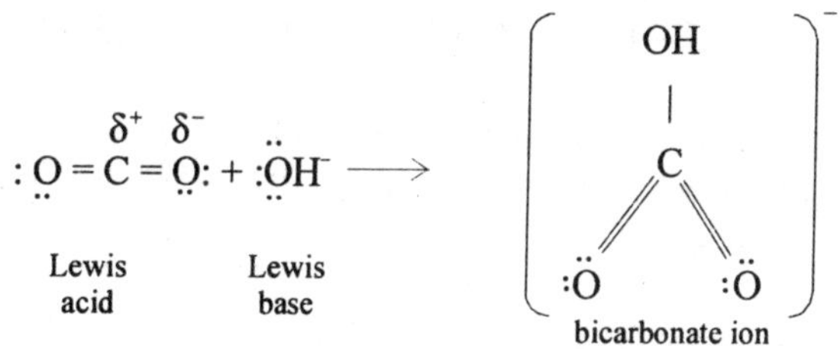

SO_2 reacts like CO_2 towards OH^-,

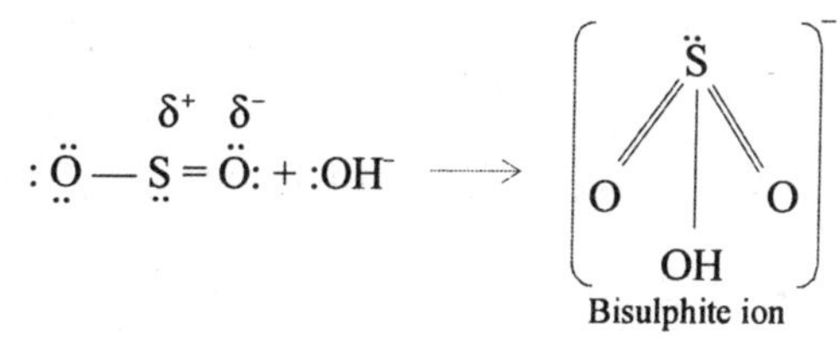

Fig. 4

5. **Elements with an electron sextet :** Oxygen and sulphur atoms contain six electrons in their valency shells and, therefore, act as Lewis acids. The oxidation of sulphite to SO_4^{2-} by oxygen, and to $S_2O_3^{2-}$ by sulphur involves acid-reactions.

$$:\ddot{\underset{..}{O}}-SO_3^{2-} \rightarrow (O \leftarrow SO_3)^2$$

$$:\ddot{\underset{..}{S}}-SO_3^{2-} \rightarrow (S \leftarrow SO_3)^2$$

Successes

(i) Lewis concept is a broader interpretation of acid- base behaviour.

(ii) This theory is not dependent on the presence of one particular element, upon any given combination of elements, upon the presence of ions, or upon the presence or absence of a solvent.

(iii) The Lewis approach is, however, of great value in case where protonic concept is not applicable.

(iv) Lewis theory is frequently employed to interpret reaction mechanisms.

Limitations

(i) In Lewis approach there is a lack of uniform scale of acid and base strength. The strength of acid and base in terms of Lewis concept is variable and dependent on the reaction considered. In this respect it is inferior to Arrhenius concept or the Lowry concept.

(ii) The concept is so broad that it is unsuited to general chemistry, in which the explanation of acid-base behaviour is most important of an acid.

CONCEPT OF POSITIVE-NEGATIVE SYSTEM

Usanovich proposed definitions of acid and bases which are most comprehensive of all acid-base theories given from time to time. According to him, an acid is any species capable of giving cations, combining with anions or electrons, or neutralising a base to yield salt. In a similar way he defined base as any species capable of giving up anions or electrons, combining with cations, or neutralising an acid to yield a salt. The Usanivich theory includes all precious acid-base definitions and in this process it has become too general. It also includes oxidation-reduction reaction as a special class of acid-base reactions.

Usanovich has placed considerable emphasis on the central atom in the compound while accounting for the acid-base character of a substance. By co-ordination unsaturation he means the ability of an atom to increase its covalence. The acid function is determined by the presence of co-ordinately unsaturated positive particles while basic function by the presence of similarly unsaturated negative particles. For instance, in SO_2 the central atom can be considered to be sulphur, and it is co-ordinately unsaturated.

Thus, it is capable of accepting an anion such O^{2-}, thereby acting as an acid. In general acidity is promoted by highly charged positive particles (Na_2O is highly basic while Al_2O_3 is less basic) and basicity by highly charged negative particle.

Acid Base Relationships on the Basis of Usanovich Concept

Acid +	Base →	Salt	Justification
Cl_2 +	2Na →	2NaCl	Na loses an electron, Cl, gains, an electron
SO_2 +	Na_2O →	Na_2SO_4	Na_2O yields O^{2-}
Sb_2S_5 +	$3(NH_4)_2S$ →	$2(NH_4)3SbS_4$	$(NH_4)_2$ S gives up S^{2-}, Sb_2S_5 gains S^{2-}

Success

Usanovich concept is useful in classifying all examples of acidity and basicity. It is most general for all other definitions of acid-base so far given.

Limitations

(i) Usanovich concept is of great value for specific problems, but for general considerations, it is undoubtedly too broad to be of wide applicability.

(ii) The theory virtually involves that all chemical reactions fall into an acid-base category, and one begins to wonder tit the purpose of using any name other than chemical reaction.

Other theories : Konopic (1949) proposed donor-acceptor acids and bases which amount to little more than that of Usanovich concept.

Bjerrum (1951) considered acid as a proton donor while a conventional Lewis acid he referred to as an antibase. Mulliken (1952) developed quantum mechanical treatment of acids and bases. But all these theories add nothing new to our knowledge.

Actually we may say that each approach is correct as far as it goes and no conflict exists around them. But one should know the main theme of each concept and should adopt his thinking to the particular problem in hand.

SOFT AND HARD ACIDS AND BASES (THE SHAB PRINCIPLE)

Based on the preferential bonding, Arland, Chatt and Davies (1958) categorised the metal ions into two classes

1. **Class (a) :** This includes ions of alkali metals, alkaline earth metals, lighter transition metals in higher oxidation states like Ti^{4+}, Cr^{4+}, Fe^{3+} and Co^{3+}, and the hydrogen ion. H^+ These ions have the following characteristic features
   - (i) They have small size, high polarising power and high oxidation state.
   - (ii) Their outer electrons or orbitals are not easily distorted

1. **Class (b) :** This includes ions of the heavier transition metals and those in lower oxidation states such as Cu^+, Ag^+, Hg_2^{2+}, Pd^{2+}, Pt^{2+}, etc. These ions have the following characteristic features
   - (i) These are large sized.
   - (ii) Their outer electrons or orbitals are easily distorted.

Depending upon their behaviour towards the metals of classes (a) and (b), the ligands have been categorised by Arland, Chatt and Davies into the following two classes:

2. **Class (a) :** This includes ligands which preferentially combine with metal ions of class (a) For example, the ligands NH_3, R_3N, H_2O and F^- ions have great tendency to co-ordinate with class (a) metal ions. The tendency of the complexation of the ligands with class (a) metal ions follows the order given below

$$F > Cl > Br > I$$

$$O > S > Se > Te$$

$$N > P > As > Sb$$

2. **Class (b) :** This includes ligands which preferentially combine with metal ions of class (b). For example, ligands such as R_3P (phosphines) and R_2S have great tendency to co-ordinate with class (b) metal ions. The tendency of the complexation of the ligands with class (b) metal ions follows the order given below:

$$F < Cl < Br < I$$

$$O \leq S < Se = Te$$

$$N \leq P < As < Sb$$

Pearson (1963) suggested that terms hard and soft can be used for class (a) and class (b) respectively. Thus, in his classification, metal ions of class (a) are called hard acids and ligands of class (a) are called hard bases. On the other hand, metal ions and ligands of class (b) are called soft acids and soft bases respectively.

Pearson's Principle : The typical Lewis acid-base reaction can be generalized as

$$\underset{\substack{\text{Lewis acid} \\ \text{(Acceptor)}}}{A} + \underset{\substack{\text{Lewis base} \\ \text{(Donor)}}}{:B} \longrightarrow \underset{\text{Complex}}{A:B}$$

To broaden the scope of the reactions we assume that the bond in AB can be ionic, polar, or non-polar. It would be very beneficial if one could predict the stability of the complex A:B. A concept designed to explain it is called the principle of Soft and Hard acids and bases and was proposed by Pearson (1963). Soft bases have donor atoms which are easily polarised and which have low electronegativity. These criteria are interdepedent, although not absolutely equivalent. They each relate to the ease with which electrons can be distorted away from the donor site.

Hard bases have properties opposite to those of the soft bases : the donor atoms tend to retain its electrons. Thus, hard bases have donor atoms with low polarizabilities and high electronegativities. Table 1 shows some typical acids classified as hard, borderline and soft. Table 1 shows some typical bases classified as hard, borderline, and soft. Within a group in the periodic table, softness increases with increasing size of the donor atom. Thus, of the halide ions., F^- is the hardest and I^- is the softest. Hard base is hard to oxidize and soft base is easy to oxidize.

Table 1 : Classification of Acids

Hard Acids	*Borderline Acids*	*Soft Acids*
H^+,Li^+,Na^+,K^+, (Rb^+,Cs^+)	Fe^{2+}, Co^{2+}, Ni^+, Cu^{2+},Zn^{2+}	$Co(CN)_5^{2-}$, Pd^{2+}, Pt^{2+}, Pt^{4+}
Be^{2+},$Be(CH_3)_2$,Mg^{3+},	Rh^{3+},Ir^{3+},Ru^{3+},	Cu^+, Ag^+, Au^+, Cd^{2+}, Hg^{2+}
Ca^{2+},Sr^{2+},Ba^{2+}	$B(CH_3)_3$,GaH_3	Hg^{2+}, CH_3Hg^+ BH_3.$Ga(CH_3)_3$,$GaCl_3$
Sc^{3+},$La^{3+}$$Ce^{4+}$,$Gd^{3+}$,$Lu^{3+}$,	R_3C^+, $C_6H_5^+$,	$GaBr_3$,$GaTl^+$, $T1(CH_3)_3$,
Th^{4+},U^{4+}, UO_2^{2+} ,Pu^{4+}	Sn^{2+}, Pb^{2+}	
Ti^{4+}, Zr^{4+}, Hf^{4+}, VO^{2+}	NO^+, Sb^{3+}, Bi^{3+}	CH_2,carbenes
Cr $^{3+}$,Cr^{6+}, MoO^{3+},WO^{4+} Mn^{2+},Mn^{7+},Fe^{3+},Co^{3+}	SO_2	π- acceptors : trinitrobenzene, chloranil, quinones, Tetracyanoe thylene, etc.
BF_3,BCl_3,$B(OR)_3$,Al^{3+},		HO^+,RO^+,RS^+,RSe^+,
$Al(CH_3)_3$. $AlCl_3$,AlH_3,		Te^{4+},RTe^+
Ga^{3+},In^{3+}		
CO_2, RCO^+, NC^+, Si^{4+}		Br_2,Br^+,I_2,I^+,ICN,etc.
Sn^{4+},CH_3Sn^{3+},(CH_3),Sn^{2+}		O, Cl, Br, I, N, RO^-, RO_2.
N^{3+}, $RPO_6{}^+$, $ROPO_2{}^+$, As^{3+}		
SO_3, RSO_2^+, $ROSO_2^+$ $C1^{3+}$,CI^{7+} I^{5+}, I^{7+}		M^+ (metal atoms) and bulk metals
HX (hydrogen-bonding molecules)		

Soft acids are those which have acceptor atoms large in size, of low positive charge and containing unshared pair (if electrons (p or d electrons) in their valence shell, *i.e.*, they have high polaritability and low electronegativity and do not have a noble gas configuration. Hard acids are those in which acceptor atoms are small in size, of high electronegativity and low polarizability and a noble gas configuration.

Soft acids form stable complexes' with bases that are highly polarisable, and are good reducing agents and not necessarily good bases towards the proton. Hard acids. On the other hand, usually form stable complexes with bases that bind well with protons.

Table 2 : Classification of Bases

Hard Bases	*Borderline Bases*	*Soft Bases*
NH_3,RNH_2,N_2H_4	$C_6H_5NH_2$, C_5H_5N,	H^-
	N_3^-, N_2	
H_2O,OH^-,O^{2-}, ROH,	NO_2^-,SO_3^{2-}	R^-,C_2H_4,C_6H_6,CN^-,RNC, CO
RO^-, R_2O		
CH_3COO^-, CO_3^{2-},NO_3^-,	Br^-	SCN^-, R_3P, $(RO)_3P$, R_3As
PO_4^{2-},SO_4^{2-},NO_2^-		R_2S, RSH,RS^-, $S_2O_3^{2-}$ I^-
F^-, (Cl^-)		

We now apply the idea of soft and hard acids and bases to the question of the stability of the complex A : B The complex AB is most stable when A and B are either both soft or both hard. The complex is least stable when one of the reactants is very hard and the other is very soft. Pearson's principle is only an approximate qualitative prediction of the relative stability for the adducts and is not a theory or any explanation of the observations.

THEORIES OF HARDNESS AND SOFTNESS

Several theories have been put forward to explain the fact that the co-ordination compounds of soft acids and soft bases are most stable. The various theories are:

1. Electronic Theory : According to the theory, hard-hard interactions involve ionic bonding and soft-soft interactions involve covalent bonding. It is expected that small sized and highly charged positive metal ions would he favouring ionic bonding with hard bases of class (a). If hard-soft interaction takes place, the resulting species would be unstable.

Misono and Coworkers (1967) proposed the following equation which correlates hardness and softness.

$$pK = -\log K = aX + bY + c$$

where K is the equilibrium constant for the dissociation of the metal-ligand complex, X,Y are the parameters of the metal ions, a and b are parameters for the ligands, and c is a constant of the ligand to adjust the pK values so that all the values lie on the same scale. For hard acids, the value of Y is found to he > 2.8 and for soft acid > 3.2. For border line acids, the value ranges between 2.8 and 3.2. For example, the values of Y are as follows:

Li^+	Al^{3+}	Mg^{2+}	Na^+	K^+	Ca^{2+}	Fe^{3+}	Co^{3+}	Cs^+	Co^{2+}
0.36	0.70	0.80	0.93	0.93	1.62	2.32	2.56	2.73	2.96

Hard acids

Sn^{2+}	Tl^{3+}	Cu^{2+}	Pb^{2+}	Tl^+	Hg^{2+}	Au^+
3.17	3.2	3.45	3.58	3.78	4.25	5.95

Soft acids

The value of b (ligand parameter) also increases as we move from a hard base to a soft base.

OH^-	NH_3	Cl^-	Br^-	F^-	$S_2O_3^{2-}$
0.40	1.08	2.49	5.5	7.17	12.4

It is interesting to note that a and X are also regarded as another measure of hardness. However, these include the inherent acid base strengths also X has been found to he closely related to the electronegativity of the metal ion and is considered to be measure of the tendency of the metals ions to accept electrons from the ligands. Further X follows the same order as that William's order for the stability constants of the complexes

Mn (II) < Fe (II) < Co (II) < Ni (II) < Cu (II)

2. π Bonding Theory : J. Chatt's pi-bonding theory can be applied to explain the bonding between the soft-soft complexes. According to this theory, the soft acids have loosely held outer d-orbitals electrons which can form pi-bonds by donating to suitable ligands. Such ligands are those in which empty d-orbitals are available on the basic atoms, *e.g.*, P, As. S, I, etc. The presence of d-orbital, on the ligands (except on CO) helps to strengthen the n-bonding.

3. Pitzer's Theory : This theory of London and van der Waal's forces explains that the forces between the interacting soft-soft groups depend upon the product of polarizabilities of the interacting groups. Both soft acids and soft base have larges values of polarizabilities.

Electronegativity and Hardness and Softness : Hardness and softness have been found to be related to electronegativity. In brief, highly electronegative species are hard acids while lower electrongative species are soft acids. For example, highly electronegative ions like Li^+, Na^+, etc. are hard acids while transition metal ions having low electronegativities like Cu^+, Ag^+, Cu^+, etc. are soft acids.

Acid and Base Strength of SHAB : Any acid or base may be considered as hard or soft by its apparent preference for the hard and soft reactants. For instance, a base B is considered to be hard or soft depending upon the value of equilibrium constant K for the following reaction:

$$BH^+ + CH_3Hg^+ \rightleftarrows Ch_3HgB^+ + H^+ \quad \text{...(1)}$$

$$BH^+ \rightarrow B + H^+$$

$$k = \frac{[CH_3HgB^+][H^+]}{[CH_3Hg^+][BH^+]} = \frac{[CH_3HgB^+]}{[CH_3][Hg^+][B]} \frac{[H^+][B]}{[BH^+]} \quad \text{...(2)}$$

$$= K_s K_a \quad \text{...(3)}$$

or $$\log K = \log K_4 + \log K_a \quad \text{...(4)}$$

If the value of K is less than unity, B is termed as a hard base and if it is greater than unity, B is termed as a soft base.

In equation (4), K, denotes the formation constant of CH_3HgB^+ from CH_3Hg^+ and B while K_a denotes the acid dissociation constant of the conjugate acid BH^+. From the values of K and K_a, it is also possible to decide whether B is hard or soft. If pK_a is more than log K_s, B is termed as hard base and if pK_a is less than log K_s, B is termed as soft base.

From the equation (2), it also follows that the reaction (1) will go to the left if B is a hard base. However, it will go to the right, if B is a soft base.

Now the question arises : Why methyl mercury (II) cation is selected in the above discussion 7 The simple reasons for this are :

(i) It is a typical soft acid, (ii) It is univalent like proton, making the calculations simpler.

Hardness and softness arc quite different from the inherent acid-base strengths because these simply refer to the hard-hard and soft-soft interactions. For example, both hydroxide ion and fluoride ion are hard bases. However, the hydroxide ion (whose pK_a is 15.7) is 10^{13} times more basic than the fluoride ion (whose pK_a, is 2.85).

Certain examples are known in which a strong acid or base may displace a weak acid or base although this may violate the principle of SHAB interactions. For example, a stronger soft base sulphite displaces a weaker hard base fluoride from a hard acid proton:

$$SO_3^{2-} + HF \rightarrow SO_3^- + F^-, K_{eq} = 10^4$$

Similarly, a very strong hard base displaces a soft weak base from a soft methyl mercury (II) cation.

$$OH^- + CH_3HgSO_3^- \rightarrow CH_3HgOH + SO_3^-, K_{eq} = 10$$

The above replacement reactions take place against the hardsoft principle. However, when both the strengths and hardness are taken into cansiderations, the hard-soft rule works well. For example, consider the following reactions:

$$CH_3HgF + HSO_3^- \rightarrow CH_3HgSO_3^- + HF\ (K_{sq} = 10^3)$$

$$CH_3HgOH + HSO_3^- \rightarrow CH_3HgSO_3 + H_2O\ (K_{sq} = 10^7)$$

In the above reactions both sulphite (pK_a = 14.2) and hydroxyl ion (pK_a = 15.7) combine with protons. But the pK_s for sulphite ions with methyl mercury (II) cation is 21.2 while in case of hydroxyl ion is only 9.4 Therefore, sulphite is considered to be a soft base.

Applications of the SHAB Principle

The principle of soft and hard acid-base finds applications in various domains of chemical reactions. Some of them are:

(i) The simplest reaction that can be explained on this concept is

$$LiI + CsF \rightarrow LiF + CsI$$

This is an exothermic reaction with 13.9 kJ/mole heat of reaction. Here soft iodide prefers to combine with the soft caesium ion. The hard fluoride ion combines with hard lithium ion.

(ii) For heterogeneous catalysis, SHAB principle says that soft metals adsorb soft bases. For solubility, the rule is that hard solutes dissolve in hard solvents and soft solutes dissolve in soft solvents.

(iii) *Occurrence of minerals* : The kinds of metal ores found in the earth's crust can be rationalized by using this principle Thus, hard acids such as Ca^{2+}, Mg^{2+} and Al^{3+} appear as $CaCO_3$, $MgCO_3$ and Al_2O_3 respectively. The anions CO_3^{2-} and O^{2-} are also hard. These three hard acid cations are never found as their sulphides since S^{2-} is a soft base. However, soft acid such as Cu^+, Ag^+ and Hg^{2+} are found to be combined with the soft base S^{2-} as sulphides. The intermediate acids such as Ni^{2+}, Pb^{2+}, and Cu^{2+} are found as both sulphides and carbonates.

(iv) In both BF_3 and BH_3, the boron is trivalent but quite different behaviour can be explained on the basis of SHAB principle. The presence of hard fluoride ions in BF_3 makes it easy to add other hard bases, eg. $H_3N \rightarrow BF^3$ the presence of soft hydride ions in BH_3 makes it easy to add other soft bases. In BF_3 the boron is nearly B^{3+} . But the soft hydride ions donate negative charges extensively to boron, so that the boron atom is almost neutral and becomes soft.

(v) For elements of variable valency there is usually an increase in hardness with increase in oxidation state. The nickel (0) in. Ni $(CO)_4$ is soft, mickel (II) is borderine and nickel (IV) is hard.

(vi) The sulphur end of the thiocyanate ion is assumed to be softer than the nitrogen end, and hence prefers soft Lewis acids.

(vii) *Relative stabilities of Compounds* : By using Tables I and II and Pearson's rule, it becomes possible to explain the relative stabilities of the compounds and complexes. For example, AgI_2^- is stable, but AgF_2^- does not exist. As described earlier, Ag^+ is a soft acid F^- is a hard base and I^- is a soft base. Therefore, AgI_2^- (soft acid + soft base) is a stable complex and can explain that CoF_6^{2-} (hard acid + hard base) is more stable than CoI_6^{2-} (hard acid + hard base).

By ucing the SHAB principle, it is possible to predict the relative strength of halogen acids in aqueous medium. For example.

$$HX + H_2O \rightarrow H_3O^+ + X$$

(where X = F , Cl , Br or I)

In the above reaction, the hard acid H+ combines with the hardest base F to form the dissociated or weakest acid. Thus, the strengths of the halogen acids will follow the following order:

$$\underset{\text{Weakest acid}}{HF} < HCI < HBr < \underset{\text{strongest acid}}{HI}$$

Out of the two complexes, $[Cd(NH_3)_4]^{2+}$ and $[Cd(CN)_4]^{2-}$, latter should be more stable (combination of soft acid Cd^{2+} with soft base (CN^-) on the basis of SHAB principle.

(iii) *Poisoning of metal catal catalysts :* This has been explained on the basis of SHAB principle. Soft metals (e.g Pd and Pt) act as catalysts and they are easily poisoned with CO, olefins, phosphorus or arsenic ligands (all soft bases). These ligands adsorb strongly on the surface of the metal and block the active sites. The soft metal catalysts are not affected by hard bases such as F.O.N.

(ix) *Rates of chemical reactions :* The SHAB principle has been used to correlate the rates of chemical reactions. The rates of electrophilic and nucleophilic reactions have been found to depend upon the hardness or softness of the acid or base.

(x) *Course of reactions :* The SHAG principle has been used to predict the course of a reaction. Consider the following reactions:

$$H^+ + CH_3HgOH \rightarrow H_2O + CH_3Hg^+ \quad ...(1)$$

$$H^+ + CH_3HgSH \rightarrow H_2S + CH_3Hg^+ \quad ...(2)$$

According to SHAB principle, the reaction (1) should move towards right because the hard acid H^+ is combining with hard base OH^- the reaction (2) should move towards left because the soft methyl mercury CH_3Hg^+ acid is combining with soft HS^- base.

LIMITATIONS OF SHAB CONCEPT

The various limitations of this concept are as follows:

(1) It does not involve any quantitative scale of measurement.

(ii) In order to explain the observed phenomenon, one has to break a compound to acid and base fragments. This breaking will be correct only if the reaction is known. For example, the break I for the scarification reactions explains the observed behaviour. However, the break 11 cannot be ignored and should be taken into consideration.

$$CH_3CO^+OH^- \rightarrow \quad C_2H_5O^-H^+ \text{ Break I}$$

$$CH_3COO^-H^+ \rightarrow \quad C_2H_5^+OH^- \text{ Break II}$$

(iii) In this concept, it is assumed that hard-soft factors do not depend upon the acidic or basic character of the compounds and therefore they work independently. However, many examples are known which show interdependence of the two concepts.

(iv) Due to the soft-soft (CH_3^+–H^-) interactions, it is expected that the following reaction should proceed.

$$CH_3^+ (g) + H_2(g) \rightarrow \quad CH_4 + H^+(g)$$

However, the unfavourable entropy change (3600 kJ $mole^{-1}$) indicates that this reaction should not proceed at all and the reason for this is that there is greater acidity of the proton relative to the CH_3 nucleus.

Conclusion

The hard-soft classification is a useful concept, but it does not lead directly to a scale of acid-base strength. Inherent acid-base strengths are also not taken into consideration. Thus the OH^- and F^- ions are both hard bases.

But the former is nearly 10^{13} times as strong a base as the latter. Various approaches to correlate inherent acid-base strength with attributed hardness (and softness) factors are yet to be widely accepted.

Interpretation of many reactions by splitting the participants into acid-base fragments is also arbitrary to some extent. The very common reaction between ethanol and acetic acid may be interpreted in either of two ways:

$$CH_3COO^-H+ + C_2H_5 + OH^- \text{ or } CH_3CO + OH^- + C_2H_5O^-H^+$$

Either interpretation is justifiable by the hard-hard combination of H^+ with OH^-. The actual involvement of the second mechanism has to be understood by other means.

Sometimes the hard-soft principle fails to keep parity with inherent acid-base strengths. The reaction:

$$CH_3^+ (g) + H2(g) \rightarrow \quad CH_4(g) + H^+(g)$$

should be favourable in view of the soft-soft combination of CH_3^+ with H^-. But actually the combination in endothermic by about 360 kJ per mole. (This has to be explained by the greater acidity of H^+ relative to CH_{3+}).

Similarly, hard-soft combinations take place in many cases like

$$SO_3^{2-} + HF \rightarrow HSO_3 + F^-$$

Here the soft bane SO_3^{2-} replaces the hard base F^- and combines with the hard acid H^+. The soft SO_3^{2-} is thus stronger than the hard F^-.

However, we should restrict our expectations in view of the statement made by Pearson. "It should be stressed that the SHAB principle is not of theory but is a statement about experimental facts. Accordingly an explanation of some observation in terms of hard and soft behaviour does not invalidate some other, theoretical explanation."

STRENGTH OF HYDRACIDS

The relative strengths of the hydracids of Group VI A and VII A have been found to be as follows:

(i)	H_2Te >	H_2Se >	H_2S >	H_2O
	3	4	7	16
(ii)	HI >	HBr >	HC1 >	HF
	–10	–9	–7	3

The above trend is also found in hydracids of Group V A. For example, the pK_a for ammonia (35) is more than that for phosphine (27).

From the above trend it appears that the strength of the acid is decreasing with the increase in the electronegativity of the elements. However, this trend is not found in the ionic character of the bonds because the ionic character has been found to increase with the electronegativity of the element.

BIOLOGICAL METALS

From the chemical view point, the living beings are having two types of constituents :

(a) **Organic constituent :** It is the main constituent. It is derived from inorganic elements (C, H,O and N). It comprises of about

90% of the solid matter, remaining 10% being the inorganic. Examples of organic constituents are proteins, carbohydrates, fats, etc.

(b) **Inorganic constituent :** Although it constitutes a relatively small amount of the total body, yet it is not less important than organic constituent for maintaining the vital activities of a living being.

When the analysis of ash of animal tissues is carried out, it is having at least 29 elements. Broadly speaking, it is possible to divide these elements into two main groups such as *essential* and *nonessential elements*.

(a) **Essential elements :** These are the elements which are indispensable for maintaining normal living state of a tissue or the whole of the body. Depending upon their absolute amounts in the body, it is further possible to divide these elements into two sub-groups :

(i) **Macro elements :** These are the elements which are required to be present in the diet in amounts more than 1mg. These elements form nearly 60–80% of all the inorganic in the body.

Macroelements include twelve elements such as carbon, hydrogen, oxygen, nitrogen, sodium, potassium, calcium, magnesium, iron, phosphorus, sulphur and chlorine. Among these the first four elements occur in substantial amounts in every body tissue and get derived from dietary carbohydrates, lipids and proteins. The body gets oxygen directly from atmosphere. Nearly 85% of the total oxygen and nearly 70% of the total hydrogen occur together in the form of water which makes nearly 3/5th of the total body weight.

The remaining amount of oxygen and hydrogen, all nitrogen, most of the carbon and some of sulphur are derived from carbohydrates, lipids and proteins which fulfil the basic requirement of tissue structure and the synthesis of various biochemical substances within these structures. Among the twelve elements, only Na, K and Cl have been described in this chapter.

(ii) **Micro elements :** These are the elements which are needed by the body in very small amounts, almost in micrograms or nanograms. These are also formed as trace elements or oligo elements. The examples of micro elements are copper, zinc, cobalt, manganese, molybdenum, iodine and fluorine. In this chapter, only zinc and cobalt have been described.

(b) **Non-essential elements :** These include the remaining ten elements which are not actually non-essential but their function in the body is not yet known. Examples of non-essential elements are bromine, boron, silicon, arsenic, nickel, aluminium, lead, tin, vanadium and titanium.

Sodium (Na), Potassium (k) and Chlorine (Cl)

As there exists intimate relations among sodium, potassium and chlorine, these should be discussed together rather than separately.

Source : The main source of sodium and chlorine in the diet is the common table salt (sodium chloride) which finds use in cooking. In general, food from plant sources is generally low in sodium. Potassium occurs in almost all foods, both plant and animal. The good sources of potassium include coffee, tea, cocoa, dried beans, molasses, most vegetables (especially green, leafy vegetables), milk, fish, chicken, beef liver, beef, pork; dried apricots, dried peaches, bananas, the juices of oranges, tangerines and pineapples, yams, winter broccoli, potatoes, and Brussels sprouts.

Healthy individual, under normal dietary conditions, seldom from a deficiency of sodium, chlorine, or potassium. However, extreme perspiration over a prolonged period, diarrhoea. or vomiting may give rise to a sodium-chloride deficiency. On the other hand, under certain health conditions, like congestive heart disease, hypertensive high blood pressure, and kidney disease, a sodium-restricted diet is preferred. As sodium in the blood and interstitial fluids is a major factor in determining the osmotic pressure of these fluids, a reduction in sodium causes the transfer of water into the cell and subsequent loss of water and sodium chloride in the urine.

Absorption : Generally, sodium, potassium, and chlorine are almost completely absorbed from the gastrointestinal tract, less than 2% of ingested sodium and less than 10% of the potassium being eliminated in the faces.

Distribution : Sodium and potassium occur in plants as well as the salts (chlorides, phosphates and carbonates) of inorganic acids and salts of proteins and organic acids. Sodium has been the main extracelluar cation, whereas potassium has been the main intracellular cation.

Functions : The metabolism of sodium, potassium, and chlorine have been intimately related to certain fundamental physiological mechanisms.

1. *Maintenance of normal hydration and osmotic pressure :* The main function of these three ions in the body has been to maintain the normal osmotic pressure of the different body fluids throughout the body and thereby to protect the body against excessive loss of fluid, which otherwise will disturb the normal hydration. But the osmotic pressure of a fluid is dependent upon the total cations of that fluid. Therefore, sodium, which is the chief cation of extracellular fluid, plays a dominant role in this connection.
2. *Maintenance of normal acid-base equilibrium :* The sodium salts and potassium salts with the corresponding weak acids are known to form the chief buffer systems respectively of extracellular and intracellular fluids. These buffer systems play vital role in the regulation of pH of various fluids under various physiological conditions.
3. *Transport of CO_2 :* All the three, mainly chloride ions are known to play role in the gaseous transport of CO_2.
4. *Neuromuscular irritability :* Both sodium and potassium play an important role in maintenance of normal neuromuscular irritability and excitability. They have been antagonistic to calcium and magnesium ions which are expressed by the following equation.

   $$\text{Irritability} \propto \frac{Na^+ + K^+}{Ca^{2+} + Mg^{2+} + H^+}$$

   Thus, the ratios of various ions as well as their absolute amounts in the body fluids and cell have been of vital significance.
5. *Maintenance of proper viscosity of blood :* In blood plasma sodium and potassium chlorides are having the outstanding function not only of keeping the globulins in physical solution

but also of regulating the degree of hydration of the plasma proteins, so important in the maintenance of the proper viscosity of blood.

6. *In secretion of digestive fluids* : Gastric HCl gets derived from NaCl of blood ; and the base in alkaline digestive juices, like the pancreatic juice and bile, get derived from blood sodium and potassium salts.
7. *In the storage of protein and glycogen* : Potassium enters the intracellular fluid during the period of cell growth and repair, i.e increased protein anabolism taking place spontaneously or induced by such agents as growth hormone or androgen. Deposition of 1 g of cell protein needs retention of about 0.4 mEq of potassium. Similarly, storage of 1 g of glycogen in the liver or muscles causes the passage of about 0.15 mEq potassium in the intracellular fluid. Because of this reason, insulin and glucose, administered into a patient of diabetic comma to enhance the glycogenesis, may bring about hypokalemia.

Excretion : Both sodium and chloride get excreted mainly by kidneys in urine, to a lesser extent, by skin in the form of perspiration and appreciably small in the digestive fluids by the gastrointestinal tract. Potassium, on the other hand, gets normally excreted almost entirely by kidneys in the urine.

Under normal physiological conditions the daily excretion of these electrolytes is almost equal to the intake. But when the conditions are not normal, the excretion of these electrolytes may get decreased or increased giving rise to hyper or hypo level in the plasma.

Difference between the Role of Na and K : In view of the close similarity of chemical properties between Na and K, it is surprising that their biological functions are very different. Na^+ are actively expelled from cells, whereas K^+ are not. This ion transport is sometimes called a sodium pump, and it involves both the active expulsion of Na^+ and the active take-up of K^+. Analysis of the fluids inside and outside animal cells shows that ion transport really does occur. In animal cells the concentration of K^+ is about 0.15 M and the concentration Na^+ is about 0.01 M. In body fluids (lymph and blood) the concentration of K^+ and Na^+ are about 0.003 M and 0.15M respectively. The transport of ions requires energy, and this is obtained by the hydrolysis of ATP.

It is estimated that hydrolysis of one ATP molecule to ADP provides enough energy to move three Na^+ ions out of the cell, and two K^+ and one H^+ ions back into the cell. The mechanism for ion transport involves polyethers natural to the organism.

The different ratio of Na^+ to K^+ inside and outside cells produces and electrical potential across the cell membrane, which is essential for the functioning of nerve and muscle cells. The movement of glucose into cells is associated with Na^+ ions; they enter the cell together. This is favoured by a high concentration gradient. The Na^+ ions entering the cell in this way must then be expelled. The movement of amino acids is similar. K^+ ions inside the cell are essential for the metabolism of glucose, the synthesis of proteins, and the activation of some enzymes.

The 1987 Nobel Prize for Chemistry was awarded to C.J. Pedersen, J.M Lehn and D. Cram for their work on the discovery and discovery and application of crown ethers and cryptates.

IRON (Fe)

Source : The main non-vegetarian dietary sources of iron include liver, kidney, heart, fish, spleen and egg yolk. The main vegetarian dietary sources include beans, peas, whole wheat, and oat meals. Other good sources are spinach, carrots, bananas, beef, cheese, tomatoes; potatoes, apples, orange, and milk. The vegetables cut by rusty knives and cooked in the ion-containers also form important sources of iron.

In the normal diet, the amount of iron gets varied form 15–20 mg/day which is sufficient in almost all cases. As women normally need more iron than men, some women do develop anemia caused by iron deficiency. In these cases the diet must be readjusted to take in more iron-rich food or the dietary iron must be supplemented by taking iron containing tonics.

Absorption : The absorption of iron is usually poor and under the normal conditions of health and nutrition, only 10–15% of the dietary iron gets absorbed in the ferrous ($Fe2^+$) from mainly in the upper duodenum, but to a lesser extent in the stomach and throughout the small intestine.

Distribution : The amount of iron in the body is appreciably small in comparison to its essential nature to the life processes. A healthy adult is having about 3.5 –4g (about 45 mg/kg body weight) of iron which occurs mainly in the form of haemoglobin; other iron containing compounds present biological system include myoglobin, cytochromes, catalase (all enzymes), etc.

Functions : Iron undertakes the following vital functions :

1. Carrier of oxygen in blood : Iron is the most important constituent of haemoglobin which gets carried oxygen from lungs to other tissues.

Haemoglobin is the red pigment of blood. It is a chromoprotein which is a type of conjugated protein. Its protein part is *globin* (94%) while the prosthetic group is iron containing complex compound called *haem or haemin.* In case the iron atom, as normally found, exists in the ferrous state the complex is *haem* which is a neutral molecule. However, if the iron atom is present in the ferric state, the complex is termed as *haemin* which carries a unit positive charge.

In the animal body, haemoglobin readily undergoes combination with oxygen to form *unstable* oxhyaemoglobin. The oxyhaemoglobin loses its oxygen in its way to various parts of the organism, thus haemoglobin acts as a carrier of oxygen from the organs of respiration to the tissues where it (oxygen) finds use in metabolic reactions. It is possible to explain the poisonous character of carbon monoxide by the fact that hamemoglobin forms a compound with carbon monoxide which is *more stable* than oxyhaemoglobin; so carbon monoxide is able to displace oxygen in the blood with the result the tissues do not obtain oxygen for their various metabolic reactions (causing death of the organism).

$$\text{Haemoglobin} + \text{CO} \rightarrow \underset{\text{(more stable than oxyhaemoglobin)}}{\text{Carboxyhaemoglobin}}$$

2. Oxygen supply to muscles : The muscles are known to store oxygen in combination with myoglobin, which is also having iron.

3. Role in biological oxidation : Iron is known to form an integral part of all the cytochromes and certain other enzymes, like catalase and peroxidases. Cytochromes are known to catalyse biological oxidation and provide energy. It has also been found that iron forms a part of prosthetic group of some of the flavoproteins and thus gets involved in electron transfer.

4. In the development of normal R.B.Cs. : The development of R.B.C takes place through a number of stages. One of these stages needs iron for the synthesis of haemoglobin. If iron is not available at this stage, the synthesis of haemoglobin will get hampered giving rise to development of abnormal erythrocytes. Thus iron is needed in sufficient amounts for the normal development of erythrocytes.

5. Relation with cell nucleus : The chromatin material of nucleus is having iron which possibly takes an essential part in metabolic oxidations occurring in nucleus.

6. Relation with oxidation in nerve cells : Iron also occurs in nissl granules of the neurons. These granules occur only in the resting state of neurons and disappear in the active state revealing the fact that iron plays an important role probably in the metabolic oxidation.

HAEMOGLOBIN

The human body contains about 4g of iron. About 70% of this is found as haemoglobin, the red pigment in the erythrocytes (red blood cells). Most of the rest is stored as ferretin. The function of haemoglobin is to pick up oxygen at the lungs. The arteries carry blood to parts of the body where oxygen is required such as the muscles. Here the oxygen is transferred to a oxyoglobin molecule, and stored until the oxygen is required to release energy from glucose sugar. When O_2 is removed from haemoglobin, it is replaced by a water molecule. Next the protein part of haemoglobin absorbs H^+. Indirectly this helps remove CO_2 from the tissues, since Co_2 is converted to HCO_3^- and H^+. The blood removes the more soluble HCO_3^- ions and the reduced haemoglobin removes the H^+. The blood returns to the heart through the veins. It is then pumped to the lungs, where the HCO_3^- ions are converted back to CO_2. This is excreted into the air in the lungs and exhaled. The blood picks up oxygen again, and the process is repeated.

The important factor in haemoglobin acting as an oxygen carrier is the reversibility of the process. If too stable an oxygen complex were formed when too much energy would be released in the lungs leaving less energy when oxygen is released in the muscles. The oxygenated from is called *oxhaemoglobin* and the reduced form is called *deoxyhaemoglobin.* This transfer of O_2 is remarkable because it involves only Fe (+II), and not Fe(+III). Other groups such as CO, CN^- and PF_3 can occupy the O_2 site. Coordination is still reversible, but is much stronger. In the case of CN^-, coordination is irreversible. These ligands reduce or prevent oxygen transfer and may result in death. However , CN^-, also interferes with the cytochrome enzyme system, which is the principal reason for its extreme toxicity.

Haemoglobin has a molecular weight of nearly 65,000 and is made up of four subunits. Each subunit comprises a porphyrin complex haem which contains Fe^{2+} bonded to four N atoms, and a globular protein called globin. The globular protein is coordinated to the Fe^{2+} in haem through a fifth N atom from a histidine molecule in the protein. The sixth position round the Fe^{2+} is occupied either by an oxygen molecule or a water molecule.

In oxyhaemoglobin the Fe^{2+} is in the low spin state and is diamagnetic. It is just the right size to fit in the hole at the centre of the porphyrin ring. The porphyrin is both planar and rigid. In deoxyhaemoglobin the Fe^{2+} is in the high-spin state and is paramagnetic. The size of Fe^{2+} increases by 28% when it changes from low spin to high spin, *i.e.,* from 0.61 Å to 0.78 Å. Thus in deoxyhaemoglobin the Fe^{2+} is too large to fit in the hole at the centre of the porphyrin, of O_2 changes the electronic arrangement of Fe^{2+} and also distorts the shape of the complex. The globular protein appears to be essential, since if it is removed oxidation of Fe(+II) to Fe (+III) occurs which is not reversible.

Haemoglobin is made up of four subunits, and when one subunit picks up an O_2 molecule, the Fe^{2+} contracts and moves into the plane of the ring. In doing so, it moves the histidine molecule attached to it and causes conformation changes in the globin chain. Since this chain is hydrogen bonded to the other three units, it changes their conformations too, and enhances their ability to attract O_2. This phenomenon is called the *cooperative effect.* The affinity of haemoglobin for O_2 decreases as the pH decreases, but as blood is well buffered this has only a slight effect. In a similar way when the blood reaches the muscles, once one O_2 has been released the others are released even more easily due to the cooperative effect in reverse.

CO_2 is the end product from the breakdown of glucose to release energy. There is an appreciable build-up of CO_2 in the muscles. This is removed from the tissue and converted into soluble HCO_3^- ions.

$$CO_2 + H_2O \rightarrow H^+ + HCO_3^-$$

This process is facilitated by the terminal amine groups of deoxyhaemoglobin which pick-up the protons produced and thus act as a buffer. The reverse process occurs at the lungs.

The porphyrin ring is conjugated and planar. The characteristic red colour arises from charge transfer between stable π and low-lying π^* orbitals on the ring and Fe.

Myoglobin

Haem is also important biologically in myoglobin which is used to store oxygen in muscles. Myoglobin is similar to one of the units is haemoglobin. It contains only one Fe atom, has a molecular weight of about 1700, and binds O_2 more strongly than haemoglobin.

Cytochromes

There are many cytochromes, which differ in slight detail, but these are broadly grouped together as cytochrome a, cytochrome c. The prosthetic group in all cytochromes comprises four haem units, and cytochromes have a molecular weight of about 12400. As in haemoglobin, Fe is bounded to four N atoms in each porphyrin ring, and the fifth site is occupied by a N atom from the associated protein. The big difference is that the sixth position is usually occupied by a S atom from an amino acid such as methionine, which is part of a protein.

Cytochromes are involved in the release of energy by oxidizing glucose with molecular O_2 in the mitochondria inside living cells. The cytochromes are reversibly between the (+ II). and (+ III) states. Cytochrumes a, b and c have slightly different reduction potentials and reactions involve all three one after the other in the order b, c, a. In this way the energy from oxidizing glucose is released gradually. The energy is stored in the form of adenosine triphosphate ATP. Which is used when required by the cell.

Table 3 : Reduction Potentials E^0 For Cytochrome

Cytochrome b	0.04 V
Cytochrome c	0.26 V
Cytochrome a	0.28 V

Ferretin

Animals, including man, absorb iron as Fe (+ II) from food in their digestive systems. Their requirements for Fe are very small as the existing Fe in the body is recycled. Any iron absorbed immediately reacts to form transferrin. A human body contains about 4g of Fe, roughly 3g as haemoglobin, and 1g as ferretin. Ferretin is found in the spleen, liver and bone narrow. When a red blood cell has become aged after an average of 16 weeks, the haemoglobin is broken down and the iron is recovered by transferrin, a non-haem protein. This is a single chain polypeptide of molecular weight 76000–80000, which contains two Fe atoms. This transports the Fe to the bone marrow where it is converted to ferretin which is brown and water soluble. Ferretin contains about 23% Fe. It consists of a roughly spherical sheath of protein called apoferretin which is approximately 120Å in diameter and which enclose a micelle of Fe(+III) hydroxide, oxide and phosphate. The micelle is an aggregate of particles whose surface bears a charge. The micelle contains 2000–5000 atoms of Fe

Catalases and Peroxidases

Catalases are enzymes which promote the disproportionation of H_2O_2. They also catalyse the oxidation of substrates by H_2O_2. The catalase molecule contains four Fe(+III) haem b groups, and has a molecular weight of about 240.000.

$$2H_2O_2 \rightarrow 2H_2O + O_2$$

The peroxidases also catalyse the decomposition of H_2O_2 but they are associated with a coenzyme AH_2, which is oxidized in the reaction.

$$H_2O_2 + AH_2 \rightarrow 2H_2O + A$$

Ferredoxins

These are a group of non-haem iron proteins which are responsible for electron transfer in plants and bacteria. They serve the same function that cytochromes perform in animals, but ferredoxins have much lower molecular weights (6000–12000), and they may contain one, two, four or eight Fe atoms. Rubedoxin is the simplest and contains only one Fe atom.

This is surrounded by four S atoms from the amino acid cysteine which is linked into protein chains and may be represented by Fe (S–cysteine)$_4$. The other ferredoxins include inorganic sulphide ions as well as cysteine S atoms, *e.g.,* $Fe_2(S_2)$ (S-cys-teine)$_4$. Since they contain inorganic sulphide ions, treatment with dilute acids liberates H_2S.

Some bacterial ferredoxins form cluster compounds. The simplest is Fe_4S_4 (S–cysteine)$_4$. The cluster may be imagined as a cube with two Fe atoms at diagonally opposite corners on the top face and two Fe atoms occupying the other two diagonally opposite corners on the bottom face. S atoms fill the four unoccupied positions. In addition, each Fe is bonded to a S atom in a cysteine molecule.

ZINC (Zn)

Source : Although zinc has been widely distributed in foods of both plant and animal origin ; oysters and certain other sea foods, liver wheat germ, yeast and lettuce have been found to be rich in zinc. Milk is having only 2 or 3 mg zinc per litre.

Absorption : Zinc proorly gets absorbed from the intestine.

Distribution : Zinc occurs in the prostate, hair, bones, liver, muscles, pancreas, gastrointestinal tract, spleen, and blood, in order of decreasing concentrations. The average concentration in the blood has been 900 mg/100 ml.,. about 75% of which has been in the red blood cells (mainly in carbonic anhydrase), 3% in the leukocytes, and remaining 22% in the plasma, bound mainly to albumin (loosely) and an α–globulin (firmly).

Biological Role of Zinc : Zinc appears to be one of the most important metals biologically. It is probably second only to iron among the heavy metals (that is excluding Na^+, Ca^{2+} and Mg^{2+}). In the past 15–20 years more than 25 zinc-containing proteins have been identified, most of them enzymes. The Zn^{2+} ion is contained in several dehydrogenases, aldolases, peptidases, phosphatases, an isomerase of yeast, a transphosphorylase, and a phospholipase, attesting to its importance in carbohydrate, lipid and protein metabolism in virtually all organisms.

Table 4 : Some Metalloenzymes of zinc

Name	*Mol.wt*	*Zn atoms Per Mol*	*Sources*
Carbonic anhydraese	–30,000	1	Erythrocytes
Carboxypeptidases	–35,000	1	Pancreas
Dehydrogenases	–85,000	≥ 2	Yeast liver
Alkaline phosphatases	89,000	4	E.coli

Table 3.4 lists a few of the enzymes known to contain zinc as an essential constituent. Among the enzymes most thoroughly studied are the following :

Carbonic anhydrase which is present in red blood cells, is involved in respiration. It speeds up the absorption of CO_2 by red blood cells in muscles and other tissues, and the reverse reaction involving the release of CO_2 in the lungs. At the same time it regulates the pH.

$$CO_2 + OH^- \rightleftarrows HCO_3^-$$

Aldol dehydrogenases from diverse species including yeast, horses and humans, catalyze the oxidation of ethanol or the reduction of acetaldehyde, using diphosphopyridine nucleotide (DPN) as a co-factor. Crystalline yeast alcohol dehydrogenase has a molecular weight of 150,000 and contains four Zn^{2+} ions and binds four DPN molecules per mole. Its structure and chemistry are not yet known in detail.

Alkaline phosphatases of bacteria catalyze hydrolysis of a variety of phosphomonoester and also act as phosphotransferases. The one from E. Coil contains four Zn^{2+} ions per units of molecular weight of 89,000. Much still remains to be learned of the chemistry, and the structure is unknown.

Carboxypeptidase which is present in the pancreatic juice, is involved in the digestion of proteins by animals, and protein metabolism in plants and animals. The enzyme catalyses the hydrolysis of the terminal peptide (amide) link at the carboxyl end of the peptide chain. The enzyme is selective. It works only when the group R in the terminal amino acid is aromatic or a branched aliphatic chain, and has the L configuration.

$$\begin{array}{ccccccccccccc}
 & R'' & & O & & R & & O & & R & & O & \\
 & | & & \| & & | & & \| & & | & & \| & \\
— & CH & — & C & —NH— & CH & — & C & —NH— & CH & — & C & —OH + H_2O \rightarrow
\end{array}$$

$$\begin{array}{ccccccccccccc}
 & R'' & & O & & R' & & O & & R & & O & \\
 & | & & \| & & | & & \| & & | & & \| & \\
— & CH & — & C & —NH— & CH & — & C & —OH + NH2— & CH & — & C & —OH
\end{array}$$

Carboxypeptidase is specific for hydrolysis of C-terminal amino acids of proteins, peptides and corresponding esters. Carboxpeptidase from beef pancreas has a molecular weight of 34,300 and contains one zinc atom per mole. Chemical and X–ray crystallographic studies show that the Zn atom is at the active site and give a moderately detailed idea of how the enzyme functions and what role is played by the metal : the Zn2+ ions is bound to the enzyme by two N and one O atom, a distorted tetrahedron being completed by a water molecule. A peptide carbonyl group displaces the water molecule when the substrate is bound, and hydrolysis of the peptide bond containing the bound carbonyl group then occurs by a series of proton transfer.

Metallothioneins are curious proteins, whose function is still unknown. They have been found in kidneys, liver and gut of horses, rabbits and humans. They have molecular weights to ca. 10,000, they contain ca. 9% by weight of metal, and nearly 1/3 of their amino acid residues are cysteines(I).

```
—NH—CH—CO—
      |
     CH2SH
      (I)
```

The metal content is mainly zinc, but an appreciable fraction is cadmium. Metallothionein is the only biological material known to contain cadmium.

Functions

1. Zinc forms an integral part of several biochemically important enzymes *e.g.,* carbonic anhydrase, liver alcohol dehydrogenase, glutamate dehydrogenase, uricase, carboxy peptidase, and kidney phosphatase.
2. Insulin is known to have zinc and in diabetes, the total amount of zinc in pancreas is reduced to half.
3. In leukemia, zinc content of leukocytes gets reduced to almost 10% of the normal amount. Temporary therapeutic amelioration of the leukemic process gets accompanied by a rise in the zinc content of these cells to normal.
4. Poor body and fur growth, and a marked decrease in the catalase activity of the kidney and liver occur in mice placed at weaning time on diets very low in zinc.

Although, no cobalt-containing complex is actually proved to the involved in oxygen metabolism, there are several that are of interest as possible models for the chemistry, especially the metal to oxygen binding, that is involved in real biological systems. Of greatest interest would be those that undergo reversible oxygenation and deoxygenation in solution. Though there are fragmentary reports of others, by far the best characterized systems are schiff base complexes such as Co(acacen) in dimethyl-formamide, pyridine, and substituted pyridines: the general reaction, which is reversible at temperatures below 0°C, is typified by the following.

The initial complex has one unpaired electron, and so also do the oxygen adducts, but as data indicate that in the latter the electron is heavily localized atoms. There is also an intense absorption band in the infra red region attributable to an O–O stretching vibration. On the basis of these and other observations the adducts have been formulated as octahedra, low-spin Co(III) complexes containing a coordinated superoxide (O_2^-) ion and having a bent Co—O—O chain. This formulation is similar to that proposed by pauling for oxyhaemoglobin. A second type of complex involves the reversible formation of oxygen bridges, Co—O—O—Co.

More recently, it has been found that certain porphyrin complexes of Co in presence of base also have reversible oxygen-binding capacity and, moreover, that these complexes can be introduced into globin to form a cobalt-substituted haemoglobin that has the same type of cooperative oxygen uptake as haemoglobin itself.

Requirements : A daily intake of 0.3 mg/kg body weight has been suggested. Due to its wide distribution of zinc in nature, this amount is easily present in the diet.

COBALT (Co)

Source and requirements : One should not know the particular sources of cobalt, because the normal average diet is having about 5 to 8mg cobalt which is many times more than the recommended daily allowance (1 to 2 mg of cyanocobalamin having 0.045 to 0.09 mg cobalt).

Absorption, excretion and distribution : Nearly 70–80% of the ingested cobalt is readily absorbed from the intestine. From isotopic studies it can be seen that about 65% of the ingested cobalt gets excreted almost completely through the kidney. The maximum amount of cobalt is found in liver which is the storage site, while only small amount is found in other tissues.

Functions : Cobalt is the important constituent of ***vitamin*** $\boldsymbol{B_{12}}$ (*cyanocobalamin*) which is needed to maintain normal bone marrow function for producing erythrocytes. Therefore, a deficiency of cobalt gives rise to limitation of vitamin B_{12} supply which may give rise to nutritional type of anemia, while an excess of cobalt gives rise to overproduction of erythrocytes causing polycythemia. The polycythemic effect may be a pharmacological one due to the action of cobalt inhibiting certain respiratory enzymes, *e.g.*, cytochrome oxidase, succinate dehydrogenase, etc.

Vitamin B_{12} is an important Co complex. The vitamin was isolated from liver after it was found that eating large quantities of raw liver was an effective treatment for pernicious anaemia. Injections of vitamin B_{12} are now used for treatment (more pleasant than eating raw liver!). Vitamin B_{12} is a coenzyme, and serves as a prosthetic group which is tightly bound to several enzymes in the body, but the precise role of vitamin B_{12} is not fully understood. Dorothy Crowfoot Hodgkin was awarded the Nobel prize for chemistry in 1964 for X–ray crystallographic work including solving the structure of this enzyme. The complex contains a Co(+ III) ion at the centre of a corrin ring system.

This is similar to the arrangement of Fe(+III) in a porphyrin ring in haemoglobin except that the corrin ring is less conjugated and rings A and D are joined directly. The Co atom is bonded to four ring N atoms. The fifth position is occupied by another N from a side chain (α–5, 6–dimethylbenzimadazole and this is also attached to the corrin ring. The sixth group which makes up the octahedron is the active site, and is occupied by a CN^- group in cyanocobalamin.

The CN^- is introduced in isolating the coenzyme, and is not present in the active form in living tissue. This position is occupied by OH^- in hydroxocobalamin, by water of by an organic group such as CH_3 (methylcobalamin). or adenosine. This shows that a metal to carbon σ bond can be formed. The cobalamins can be reduced from Co^{III} to Co^{II} and Co^{I} in neutral or alkaline solutions both in the laboratory and in vivo (in the living body). The Co^{I} complex is strongly reducing. Its structure is five-coorrdinate, *i.e.,* the site usually occupied by CN^- or OH^- is vacant.

Methylcobalammn is important in the metabolism of certain bacteria which produce methane. These bacteria can also transfer a methyl group CH_3 to a few metals such as Pt^{II}, Au^{I} and Hg^{II}. The latter poses a considerable ecological problem as the bacteria can transform elemental Hg or inorganic Hg salts into highly toxic methyl CH_3Hg^+ or dimethyl mercury $(CH_3)^2$ at the bottom of lakes.

Cobalt is also biologically important in some enzymes. Glutamic mutase is involved in the metabolism of amino acids and ribonucleotide reductase in the biosynthesis of DNA. Traces of cobalt are essential in the diet of animals. Some sheep raised in Australia, New Zealand, Florida and Britain suffered from a deficiency disease which was traced to them grazing on cobalt deficient soil. This can be remedied either by treating the soil periodically, of by forcing the animals to swallow a pellet of cobalt.

This pellet remains in the rumen, and slowly releases cobalt into the gut. (Sometimes the animals are made to swallow a metal screw as well. This too remains in the rumen, and its purpose is to scrape any coating off the cobalt pellet. The pellet is recovered and reused when the animals are slaughtered).

Larger amounts of cobalt appear to be harmful. Traces of cobalt (1-1.5 ppm) are added to beer to make its froth better. This has been linked with an increased rate of heart failure among heavy beer drinkers who have a dietary deficiency of protein or thiamine.

5

Toxic Metals and Their Toxicity

INTRODUCTION

All the heavy metals have been cumulative and potentially toxic. They bring about damage to various organs such as liver, kidney and the gstroinstestinal tract. The toxic character of metals if attributed either due to (a) their combination with the sulphydryl (–SH) groups of essential enzymes to form ring compounds and thus inactivating the enzymes or (b) due to complex formation with amino, imidazole, phosphate and carboxyl groups. The common toxic metals include arsenic, lead, mercury, antimony, silver, gold, copper, iron etc. However, here only the first three toxic metals and toxicity produced by them have been described.

Arsenic

Arsenic is well known for its toxicity, because of its association with criminal and wartime use. It is widely distributed in the earth, 1.5–3.0 ppm in sedimentary rocks. Of the two oxidation states, As (III) and As (V), the latter is the more common. Arsenic is used mainly in agriculture chemicals are herbicides, insecticides, desiccants, wood preservatives and feed additives.

Arsenic has some similar toxic properties to Pb, Hg and Cd, as regards bonding to sulphur and inhibiting enzyme action such as pyruvate dehydrogenase. The order of toxicity of arsenic compounds is ; arsines (As III) > arsenite [As (III) > arsenate [As (V)] and arsenic-organic acid [As (V)]. Arsenic, which is found mainly in the liver, kidneys, lungs and intestinal walls, is readily absorbed if water soluble. Arsenic is also toxic by replacement of phosphorus in ATP and in coagulation of proteins. Since P and As are in the same periodic group, and since P is a major essential element, it is not surprising that arsenic interferes with phosphorus metabolism. In the biosynthesis of ATP, in the presence of arsenite, the intermediate 1, 3 diphosphoglycerate is not formed, but instead 1 arseno, 3-phosphoglycerate ($O_3PO–CH_2–CH(OH) – CO – OAsO_3)^{4-}$. This material hydrolyes to arsenate and 3-phosphoglycerate rather than undergoing the normal oxidative phosphorylation.

A lethal dose of the more toxic forms of arsenic is 125 mg per kg body weight, but for less toxic forms, say arsenic in drinking water, larger quantities are necessary, *i.e.,* 100–200 mg per kg body weight. The symptom of acute arsenic poisoning is severe gastro-enteritis, and in chronic poisoning, loss of weight and hair, and skin lesions. An antidote for arsenic is BAL, British Anti Lewisite.

The toxic action of As (III) is by attacking –SH groups of an enzyme, thereby inhibiting enzyme action.

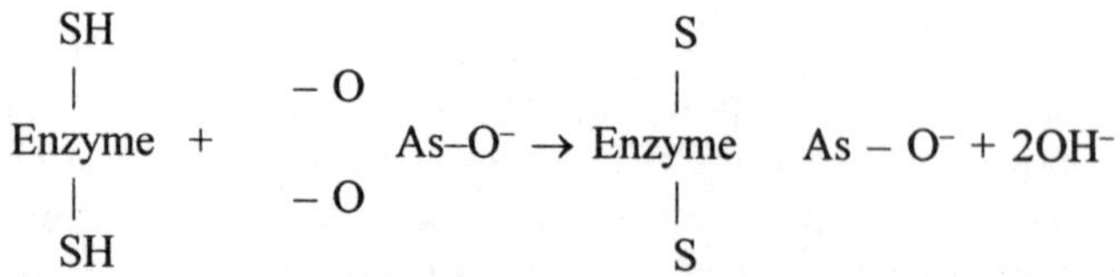

The enzymes producing cellular energy in the citric acid cycle are adversely influenced. The inhibitory action has been based on inactivation of pyruvate dehydrogenase by complexation with As (III), whereby the generation of ATP is disallowed.

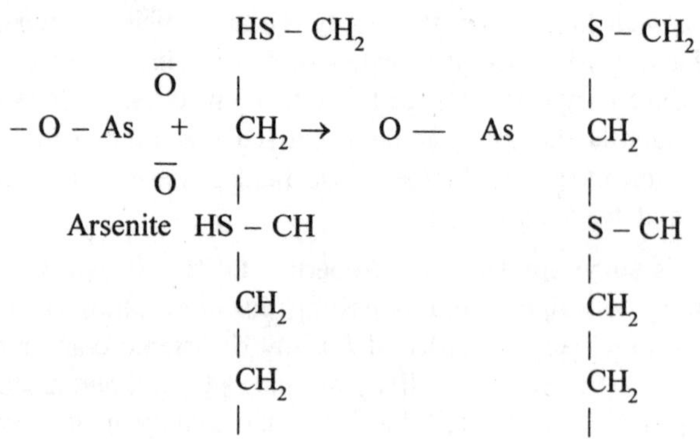

CH_2	CH_2
|	|
CH_2	CH_2
|	|
C = O	C = O
|	|
Protein	Protein
Dihydrolipoic acid-protein	Inactivated Protein complex with As(III)

Because of its chemical similarity to P, As is known to interfere with some biochemical processes which are involving P. This has been reported in the biochemical generation of the key energy yielding substance, ATP (adenosine triphosphate).

An important step in ATP formation involves the enzymatic synthesis of 1,3 diphosphoglycerate from glyceraldehyde 3-phosphate. Arsenite is known to interfere by forming 1-arseno 3-phosphoglycerate instead of 1, 3 -diphosphoglycerate. Phosphorylation gets replaced by arsenolysis which involves sponteneous hydrolysis to 3-phosphoglycerate and arsenate. Arsenic (III) compounds at higher concentrations coagulate proteins, probably by attacking the sulphur bonds retaining the secondary and tertiary structures of proteins.

The three main biochemical actions of As include coagulation of proteins, complexation with coenzymes and uncoupling of phosphorylation. The main antidotes for As poisoning include chemicals which are having—SH groups and capable of bonding to As(III), *e.g.*, 2, 3-dimercaptopropanol (BAL).

HSCH2–CH – CH2OH
　　　　|
　　　SH

2. Lead

Though it is not considered as toxic as mercury and cadmium, lead is so widespread in our consumer society, it is probably the most serious toxic metal.

The main uses of lead accumulator, tetraalkyl lead compounds, cable sheathing, lead pigments, alloys and in manufacturing–all around 9–11% each. The uses of lead most commonly associated with poisoning are lead glazed pottery, solder plumbing, in paint and an additive to petrol.

Tetraethyl lead is produced in large amounts and used an 'anti-knock' additive to increase the octane number of petrol. The commercial preparation uses a sodium/lead alloy.

$$Na/Pb + 4EtCl \rightarrow PbEt_4 + 4NaCl$$

Lead is poisonous to man, and burning petrol containing PbEt4 releases lead into the atmosphere. Its use is declining sharply due to legislation that new cars must run on lead-free petrol.

Numerous attempts have been made to estimate the lead consumption per person per day in an urban environment. One important feature is the significant amount of lead assimilated into the body from air intakes. This is because of the high absorption factor of 0.4, compared with 0.1 for food and water.

For people living close to busy roaαs, where the average lead in the air could be 2 or 3 times higher, the contribution from air intake can become quite significant (up to 50%.). An air lead level of 1 μg m^{-3} is said to raise the blood lead level by 2 μg 100 ml^{-1}. The data for children is even more tentative because of their wide range of intakes of food, water in air and variations with age.

Also children have a greater metabolism rate and absorption factor. The limit suggested by WHO for a tolerable intake of lead per day for an adult is 430 μg. No figure has been given for a child, but a value of 200 μg day^{-1} seems reasonable.

Around 70 to 90% of the lead assimilated goes into the bones, the next major accumulator is the liver, then the kidneys. Divalent lead readily replaces Ca^{2+} in bone, and either becomes firmly fixed or reversibly fixed. Lead in the latter form may be released into the blood stream and a person who has been purged of lead may have a recurrence of lead poisoning at a later date due to release from bone tissue.

The principal indicator of lead levels in human beings in lead in blood, expressed as μg 100 ml^{-1} or μ mol l^{-1}, (80 μg 100 ml^{-1} = 3.86 μ mol l^{-1}). The lead concentration can be given for whole blood, or in the red cells (paked cell volume), where it mostly resides. The latter is more useful, because the red cell level may be low if the subject has anaemia. Blood lead levels and some related health effects are given in Table 39.3. While the occupational blood lead level is around 70–80 μg 100 ml^{-1}, considerable dispute exists as to what is considered a high blood lead level for the general population. People in urban areas have levels around 10–20 μg 100 ml^{-1}, but if levels reach 30–40 μg 100 ml^{-1}, especially in children, there is reason to be concerned about the lead intake. Symptoms of lead poisoning are many and not always the same for every person. The onset of brain damage appears to be associated with high levels of blood lead, > 80μg 100 ml^{-1},. Young children are particularly susceptible, the damage showing up in behavioural and educational abnormalities, and may be mental retardation. In mild form, lead effects the central nervous system, manifested in hypertension, irritability and impaired motor skills. It is in this latter area where the debate has been fiercest.

There are many who say that there is little or no effect with blood levels around 30 μg 100 ml^{-1}, and there are many who say there is detrimental effects. From the information in Table 3.6 it is clear that even "normal" levels (<20 μg 100 ml^{-1}), affect the human biochemistry, and so lead cannot be taken lightly. The effect of lead on the renal system may be reversed if caught in time, otherwise irreversible damage may be done to the kidneys.

Alkyl lead compounds are not as common in the environment as inorganic lead. Being lipid soluble their main affect is on the central nervous system. It is thought that principal toxic metabolite is R_3Pb^+ which is produced in the liver.

The major biochemical effect of Pb has been attributed to its interference with heme synthesis, which gives rise to hematological damage. Pb is reported to inhibit several of the key enzymes which are involved in the overall process of heme synthesis whereby the metabolic intermediates get accumulated. One such intermediate has been delta-amino levulinic acid. An important phase of heme synthesis involves the conversion of delta-aminolevulinic acid to porphobilinogen :

$$HO_2C - CH_2 - CH_2 - \overset{\overset{O}{\|}}{C} - \underset{\underset{NH_2}{|}}{\overset{\overset{H}{|}}{C}} - CO_2H$$

delta–aminolevulinic acid
(ALA dehydrase : cytoplasm)

(a)

$$HO_2C - CH_2 - C - C - CH_2 - CO_2H$$

‖ ‖
C C
$H_2N - CH_2$ N H
|
H

prophobillinogen

(b)

As Pb inhibits the ALA-dehydrase enzyme it cannot react further to form (b) porphobillinogen. The overall effect involves the disruption of the synthesis of haemoglobin as well as other respiratory pigments, like cytochromes, which need heme. Finally, Pb does not àllow the using of O_2 and glucose for life-sustaining energy production. It is possible to detect this interference at a lead level in the blood of about 0.3 ppm.

The detection of (a) has been regarded as a sensitive test for Pb in the body. At higher levels of Pb in the blood (> 0.8 ppm) symptoms of anaemia due to the deficiency of haemoglobin appear. Elevated Pb levels (> 0.5–0.8 ppm) in the blood may bring about kidney dysfunction and finally cause brain damage. Because of the chemical analogy of Pb^{2+} and Ca^{2+} bones may act as repositories for Pb accumulated by the body. Furthermore, this Pb may get remobilized along with phosphates from the bones which may be able to exert a toxic effect when it is transported to soft tissue.

Methods for removal of lead from the body all depend, to some extent, on coordination of the metal to give a soluble species that may be excreted from the body. The most common method used today is chelation using $CaNa_2$ *EDTA*. Other approaches are to administer a low calcium high phosphate diet, or to prescribe sodium citrate and 2, 3-dimercaptopropanol (British Antitewisite, BAL). BAL was developed as an antidote for arsenic, but it also works for lead.

$$2HSCH_2–CH–CH_2OH + Pb^{2+} \rightleftarrows + 4H^+$$

It may be necessary for a person to have a number of removal treatments, because of continued release of lead from the bones. It is also necessary to watch for removal of other materials, such as essential trace elements by the reagents.

Two other typical Pb chelates used in the removal of lead from the body are shown below :

H_2C – – $CHCH_2$ OH OH

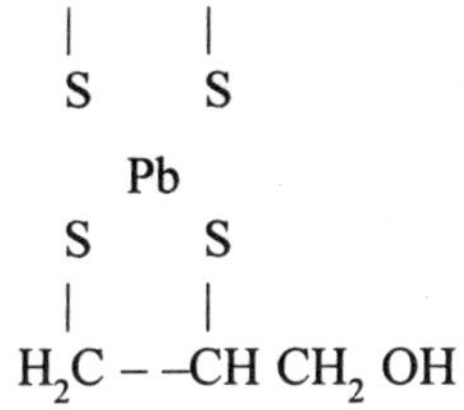

Pb – BAL chelate
(BAL = 2, 3
dimercaptopropanol)
BAL, abbrev of
British anti–lewistie

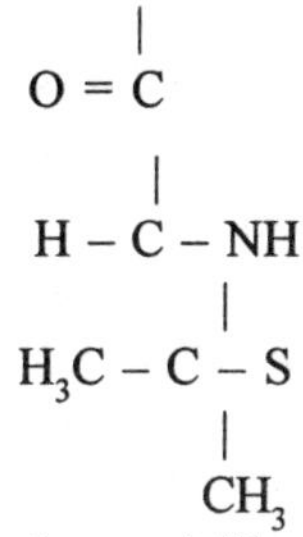

Pb – d – penicillamine chelate

Mercury

Inorganic mercurial compounds, were previously used in syphilotherapy and as laxatives. However, they are no longer used in therapeutics. The organic mercurials are being still used as antiseptics, preservatives, spermicides and diuretics.

Inorganic compounds such s $HgCl_2$. Hg_2Cl_2 and HgO are poisonous if eaten. Mercury is a cumulative poison (like Cd, Sn, Pb and Sb). Because they have no biological function, there is no mechanism for excreting them from the body. HgII inhibits enzymes, particularly those containing thiol groups.

These inorganic mercury compounds have been used for controlling the growth of slime moulds in wood pulp and paper making factories. They are also used in antifouling paints for boats. As a fungicide for treating seeds and plants, and for treating a plant disease called clubroot in brassicas.

More recently organomercury compounds such as diethyl mercury $Hg(Me)_2$ and MeHgX have been used for treating seeds against attack by fungi. Alkyl mercury compounds such as dimethyl mercury are much more toxic than inorganic mercury compounds. Aryl mercury compounds are even more dangerous. They cause brain damage giving numbness, loss of vision, deafness, madness and death.

There have been several alarming incidents involving mercury poisoning.

1. In the early 1900s mercury salts were used in the production of felt has. The dust affected the central nervous system of workers, causing tremor of muscles called 'hatter's shakes's. This led to the expression 'mad as a hatter'!
2. 52 people died at Minamata (Japan) in 1952 through eating fish contaminated by mercury. The Hg came from losses from a factory using H^{II} salts to catalyse the production of ethanal (acetaldehyde) from ethyne (acetylene). The $HgCl_2$ was converted to an organomercurial compound MeHgSMe by anaerobic bacteria in the mud on the sea bed. This is concentrated in the food chain. First it is taken up by plankton, which is eaten by the fish, and other seafood, which in turn is eaten by man. There were outbreaks of mercury poisoning in Japan in 1960 and again in 1965 arising through eating contaminated shellfish.
3. Corn seeds which has been treated with an organomercurial compound to prevent fungal attack were eaten as food rather than being planted to grow a crop in Iraq in 1956 and again in 1960. This resulted in many deaths. There have also been problems through loss of mercury salts used as antifungicides in paper mills in Sweden.

The problems of mercury poisoning are now better understood. Though inorganic mercury salts are poisonous, they are nothing like as toxic as alkyl and aryl mercury compounds. However, the loss of inorganic mercury compounds must be prevented as Hg^{2+} and Hgt_2^{2+} can be methylated by bacteria present in rivers, lakes and the sea. Algae, molluscs and fish may concentrate the small amount present by a factor of 2–10 times. and they may be eaten by something else and concentrated again.

$HgMe_2$ and $Hg - Me^+$ are both very stable and persist for a long time because they have a strong Hg –C bond. Considerable efforts are now being made to prevent the discharge of mercury compounds with industrial effluent. The main problems come from acetaldehyde and vinyl chloride monomer plants where mercury compounds are used as a catalyst, and from the electrolytic production of NaOH and Cl2 where it is used as the cathode.

4. The most recent problem is from the widespread use of mercury to extract gold in Brazil. River water and silt are passed over mercury, when the gold dissolves in the mercury, forming an amalgam. Losses of mercury have poisoned considerable stretches of the Amazon river.

MERCURY POISONING

(i) Acute mercury poisoning : Generally, it takes place due to accidental ingestion of mercuric chloride and other readily ionizable mercury salts. This poisoning is characterised by the following symptoms:

(a) *Severe gastrointestinal irritaion takes place leading to electrolyte disturbances and peripheral circulatory collapse.*

(b) *Metallic taste comes in mouth.*

(c) *Excessive salivation and inflammation of gums takes place.*

(d) *Marked renal impairment takes place at the later stage.*

It is possible to treat acute mercurial poisoning by the following methods:

(a) *By the administration of proteins in the form of raw eggs or milk with give rise to non-toxic proteinates.*

(b) *By giving suspension of medicinal charcoal.*

(c) *By gastric lavage which is to be given immediately, i.e., up to 15 minutes of ingestion because mercury gets rapidly absorbed by the mucous membrane of the alimentary tract.*

(d) *By administering dimercaprol within first hour or two after poisoning.*

(e) *By administering acetyl D-penicillamine. It has been somewhat less toxic but weaker chelating agent.*

(ii) Chronic mercury poisoning : It gets manifested by increased irritability, tremores, headache, easy fatiguability, colitis, stomatitis, increased salivation, dermatitis, electrolyte imbalance, liver damage and renal damage.

Dimercaprol has been reported to be an effective antidote.

Systemic absorption of topically applied mercurial antiseptics gives rise to an allergic manifestation which is known as *pink disease* or acrodynia. It has been found to be more common in children and its symptoms include marked swelling and erythema of the extremities. It responds to acetyl penicillamine therapy. Fever and skin rashes of various types also occur in adults using mercurial antiseptics topically.

Cadmium

It is strange that Zn in an essential element for life, but that the other two elements in the same group. Cd and Hg, are both extremely toxic. The main threat from Cd is in places near Zn smelters, as Cd may escape as dust with flue gases. The manufacture of Ni, Cd batteries has caused problems in Sweden and Japan. There is also concern at the amount of Cd in cigarette smoke. If Cd is ingested it accumulates in the kidneys. It causes malfunction of the kidneys and also replaces Zn in some enzymes, thus preventing them from working.

Mercury vapour is toxic and if inhaled can cause giddiness. tremores, lung damage and brain damage. In the laboratory. mercury should be covered with oil or toluene, and spillages should be treated with flowers of sulphur, forming HgS.

ACTINIDES (5F-BLOCK ELEMENTS)

These are the elements in which the extra electron enters 5*f*-orbitals of (*n*–2)*th main shell. These are known as 5f- block elements* and are also known as actinones. According to this definition of actinides only *thirteen* elements from Th_{90} ($5f^0 6d^2 7s^2$) to No_{102} to Lw ($5f^{14} 6d^0 7s^2$) are regarded as the members of actinide series, because all these fifteen elements are having similar physical and chemical properties. In fact the name actinides has been derived from actinium which is *prototype* of actinides as lanthanum is the prototype of lanthanides.

POSITION OF ACTINIDES IN THE PERIODIC TABLE

It is possible to discuss the study of the position of the actinides in the periodic table under the following two heads.

(i) **Prior to the discovery of the trans-uranium elements** (*i.e.*, before 1940). The existence of the lanthanide series made the discoverers to predict another series of elements resulting from the addition of electrons to an (n–2) f-shell (*i.e.*, 5f-shell)some where in the heavy elements region. Before the discovery of transuranium elements, the naturally occurring heaviest known element such as Th_{90}, Pa_{91}, and U_{92} were kept below Hf_{72}, Ta_{73} and W_{74} in IV B, VB and VIB groups of the periodic table, as these elements showed +4, +5 and +6 oxidation states and resembled Hf, Ta and W respectively in many of their properties.

(ii) **Following the discovery of the trans-uranium elements :** In 1940, the discovery of the element neptunium (NP_{93}) was reported and this discovery was followed shortly by the discovery of plutonium (Pu_{94}) in 1941. The tracer chemical experiments conducted with Np_{93} and Pu_{94} revealed that the chemical properties of these tow elements very much resemble those of U_{92}, and not at all those of Re_{75} and Os_{76}. On this basis in 1944, the position of U_{92}, NP_{93} and Pu_{94} was shown in the periodic table as provided. It was assumed that the undiscovered elements with atomic number 95 and 96 should be very much similar to that of U_{92}, NP_{93} and Pu_{94} in their chemical properties. This was proved wrong, because the experiments of discovery of NP_{93} and Pu_{94} got failed. Later in the same year (1944) Seaborg claimed that all the known elements heavier than Ac_{89} were wrongly.

GENERAL PROPERTIES OF ACTINIDES AND THEIR COMPARISON WITH LANTHANIDES

The important properties of actinides are given below.

(1) Electronic configuration and nature of bonding in actinide compounds

According to recent view 5*f*-orbitals in actinides start filling from protactinium (Pa_{91}) and not from thorium (Th_{90}) as can be seen from the valence-shell from the valence-shell electronic configuration of actinides included in Table 4.2. For the sake of comparison the valence-shell configurations of lanthanides are also Fig. 4.8. When the electronic configurations of lanthanides and actinides are compared, it can be seen that both show a substantial overall similarity and enhanced tendency for the lighter members namely Ac_{89} to Np_{93} to retain 6*d* electrons.

The latter tendency arises both from the even smaller energy separation between the 5*f* and 6*d* or orbitals in actinides than that between the 4*f* and 5*d* orbitals in lanthanides and from the difference in change in binding energy of the weakest bound (*i.e.*, most readily removable) 6*d* and 5*f* elections in actinides with atomic number as depicted which a rough quantitative picture for the binding energies of the weakest bound 5*f* and 6*d* electrons in actinides has been depicted.

Spectroscopic chemical and other data reveal that 5*f*-level becomes progressively lower in energy compared to the 6*d*-level with increasing atomic number. The relative point of change would be thus different in the two series and also the fall in energy and shrinkage in size are not so precipitous for 5*f* orbitals as for 4*f* orbitals. Therefore, the spatial extension of 5*f* orbitals relative of 6*s* and 6*p* orbitals is greater than of 4*f* orbitals relative to 5*s* and 5*p* orbitals. The greater spatial extension of *f* orbitals is also shown experimentally.

As *f* shell is being filled in lanthanide (4*f* shell) and actinides (5*f* shell), the elements of these two series have close relationship in properties with one another. Hence Ac_{89} and La_{57} occupy corresponding positions as prototypes for the two series.

Hence, 5*f*-orbitals have been more penetrating than 4*f*-orbitals. In lanthanides 4*f*-orbitals have been found to be quite within 6*s* and 5*p* orbital. Hence there is no question of an overlap between 4*f* and the surrounding ligands. The 5*f*-orbitals which have been more penetrating than 4*f*-orbitals have been found to be more diffuse at the periphery of the atom. Therefore these orbitals are somewhat disturbed by ligands.

The energy difference between different levels and sub-levels gets decreased with increasing atomic number. Hence in the actinides the energies of 5*f*, 6*d*, 7*s* and 7*p* orbitals are almost equal. This causes that unlike *lanthanide, the bonding in the compounds of actinides may be able to involve any or all of these orbitals.* This can be seen from the fact that, while lanthanides virtually form no complexes or covalent compounds in which 4*f*-orbitals participate in bonding (*lanthanides form mostly ionic compound), the* actinides are able to form complexes not only with anions such as X–, $SO^{2-}{}_4$ etc., but also with p-bonding ligands like alkyl phosphines, thioethers and chyclopentadience by the use of 5*f*-orbitals.

OXIDATION STATES AND OXIDATION POTENTIAL

When the oxidation states of lanthanides are compared with those of actinides, we find that +3 oxidation state is the most common for both the series of elements. This oxidation state would become increasingly more stable as the atomic number gets increased in the actinide series. The increasing stability of +3 oxidation state can be seen from the increasing difficulty of oxidation above +3 oxidation state. This is clearly evident from of oxidation potentials (E° values) given in the same Table. E° values have been recorded in 1M per chloric acid at 25°C. The standard electrode potentials for the lanthanides couple, Ln^o/Ln^{3+} *have been found to become steadily more positive with the increase of atomic number (due to lanthanide contraction),* while for the actinides couple An^o/An^{3+} these have been found to become more positive from Ac to U and have been found to become less positive from Ac to U *and have been found to become less positive till* Am. Am^{4+} is thus a more powerful oxidising agent than Ce^{4+}, when Pu^{4+}, Np^{4+} and U^{4+} are less powerful. U^{3+} have been a strong reducing agent.

Actinides exhibit a greater multiplicity of oxidation states. In the first half of the actinide series (*i.e.,* lower actinides), the energy needed for the conversion $4f \rightarrow 5d$. Hence, the lower actinides should, on a comparative basis, show more higher oxidation states like +4, +5, +6 and +7. Correspondingly, because in the second half of the actinide series (*i.e.,* higher actinides), the energy needed for the conversion $5f \rightarrow 6d$ has been more than that needed for the conversion $4f \rightarrow 5d$, higher actinides should exhibit more lower oxidation states such as +2.

The tripositive oxidation state is quite common in each series. The two groups of elements have been wholly comparable in this respect. The +3 state characteristic of lanthanides is not seen in aqueous solution of Th and Pa and this oxidation state becomes the predominantly stable oxidation state in aqueous solution of the actinides only when Am is reached. The elements from U to Am are having several oxidation states, while there exists no analogous example among lanthanides. These differences are attributed to the assumption that 7*s*-, 6*d*- and 5*f*- levels are having almost the same energies in this region.

As 5*f*-electrons can be more easily removed than the 4*f*-electrons, for the actinide metals the lower oxidation states are less important while the higher ions are more important compared to the lanthanide metal; *e.g.,* even Cm gives rise to Cm(IV) oxidation state in solids CmO_2 and CmF_4; while only Gd(III) compounds are known. Cm^{4+} ion has ben also stable in solution.

+ 2 Oxidation state : Only americium (analogous to europium) forms a stable +2 state. This state has been stable in CaF_2 only and has been confirmed by optical and electron spin resonance spectra. +2 state is uncommon for other actinides. An^{2+} ions have been found to resemble Ln^{2+} ions in their general chemistry.

+ 3 Oxidation state : +3 state is a general oxidation state for most of the actinides. For Th and Pa, +4 and +5 states respectively has been important. An^{4+} ions resemble Ln^{4+} ions in their properties. Many isomorphous salts are given by the elements of both the series. Trichlorides and trifluorides of Ac, U, Np, K, Pu and Am are isomorphous. On hydrolysis all the halides yield oxyhalides.

Ac, Pu and heavier elements are known to form the oxides of An_2O_3 type which are isomorphous with Ln_2O_3 oxides.

Nitrates, perchlorates and sulphates are soluble whereas hydroxides, fluorides and carbonates are insoluble. Like lanthanides, actinides are also known to form complexes with EDTA. Lanthanide- EDTA complexes are somewhat less than the corresponding actinide-EDTA complexes.

+ 4 Oxidation state : This is the main oxidation state for Th and is a stable oxidation state upto Am. Am^{4+} and cm^{4+} are known to form only complexes in concentrated fluoride solution of low acidity. General Chemistry of An^{4+} ions has been found to be similar to that of Ln^{4+} ions. The hydrated fluorides and phosphates of both An^{+4} and Ln^{4+} ions have been insoluble. ThO_2, PaO_2, UO_2, NpO_2, AmO_2, CmO_2 and BkO_2 are having fluorite structure.

The tetrachlorides and tetrachlorides of Th, Pa, U and Np are well known characterised compounds while tetraiodides of Th, Pa and U have been isolated. Oxyhalides of Th, U and Np can be obtained by heating AmX_4 with Sb_2O_3. An^{4+} ions are known to form complexes with anionic ligands like HSO^-_4, NO^-_3, Cl^- etc.

+ 5 Oxidation state : This state is quite important for Pa. Pa^{5+} resembles very much Nb^{5+} and Ta^{5+}. U, Np, Pu and Am also exhibit +5 state, but these are less characterised. The only pentahalides are found for those of Pa^{5+} and U^{5+}.

Fluoroanions of Pa, U, Np and Pu of the types AnF^{1-}_6, AnF^{2-}_7, and AnF^{3-}_8 exist in the solid state. Oxychlorides, $AnOCl_3$ (An = Pa, U and Np) are also known.

AnO^{2+} has been the most important ion which is having An^{5+} cation, It is having linear structure both in solid and solution. This monovalent dioxo cation is known to form many complexes with anionic and neutral ligands.

+ 6 Oxidation state : U, Np, Pu and Am show +6 oxidation state in divalent dioxo cation, AnO^{2+}_2. This cation has been linear both in solid and solution. The simple molecular halide, UP_2F_2 is having the linear O–U–O group with fluorine bridges. The O–U bond distance is 1.75 to 2.00 Å. The overall structure is a *flattened octahedron.* Although AnO^{2+}_2 cation is linear in shape, it is known to from complexes with exceptional geometries, *e.g.,* four, five and six coordinated complexes are given by this cation.

ATOMIC AND IONIC RADII-ACTINIDE CONTRACTION

Crystal structure data form the basis for ionic radii of the actinide elements. In Table 4.4 atomic and ionic radii of the tripositive actinide ions (*i.e.,* An^{3+} and An^{4+} ions) are given. The atomic radii of lanthanide atoms and ionic radii of Ln^{2+}, Ln^{3+} and Ln^{4+} ions are also given in the same table for comparison. It is to be noted that for An^{3+} and An^{4+} ions there exists as *actinide contraction (i.e.,* decrease in ionic radii of An^{3+} and An^{4+} ions), *analogous to the lanthanide contraction,* with increasing positive charge on the nucleus.

This actinide contraction is ascribed to the addition of successive electrons to an inner *f*-shell (*i.e.,* 5*f*-shell), so that the imperfect screening of the increasing nuclear charge by the additional 5*f* electrons causes a contraction of the outer or valence orbital. The jumps in contraction between the consecutive elements in the actinides have been found to be greater than in the lanthanides. The greater contraction *is attributed to the lesser shielding of 5f-electrons which are therefore, puller more strongly by the nucleus.*

The actinide contraction gives rise to a decrease in the basicity of the elements with increasing atomic number.

IDENTIFICATION AND SYNTHESIS OF TRANURANIUM ELEMENTS

Among the naturally occurring elements, uranium is having the highest atomic number equal to 92. Due to this reason this element continued to occupy the last position in the periodic table for a pretty long time. After 1940 fourteen elements with atomic numbers equal to 93, 94, 95, 96, 97, 98, 99, 100, 101, 102, 103, 104, 105 and 106 respectively were identified and synthesised by the transformation of naturally occurring elements by nuclear reactions. These man-made thirteen elements have been kept beyond uranium in the periodic table and are collectively called *transuranium elements.*

The year identification of the first isotope, the names of the discoverers, the method of identification and synthesis of these transuranium elements have been included in Table 4.1.

SEPARATION OFACTINIDE ELEMENTS

Transuranium elements, produced by nuclear reactions, could be isolated from the target materials and irradiated nuclear fuels by the following methods:

(1) Solvent extraction method : This method is based on the various oxidation states of actinide elements. This technique is mainly used in the recovery of U and Pu from used-up nuclear fuels. This method is based on the distribution of a metal between the aqueous solution and an organic solvent. When treated with *methyl isobutyl (hexone)* Np^{4+}, Pu^{5+}, Pu^{6+} and U^{6+} are extracted . *Diethyl ether* and *tri-nbutyl phosphate* (TBP) a other solvents which find use as extractants. Due to the high viscosity and density, TBP find use as 20% solution in kerosene. The method is preferentially applied no nitrate systems, since other ions such as sulphate, perchlorate, fluoride etc. are strongly complexing and tend to retain the metal in aqueous solution. Hexane and diethyl ether need a high concentration of NO^-_3 ions in the aqueous phase and it is achieved by adding Al $(NO_3)_3$ which is having a high salt-out action. TBP is resistant to nitric acid oxidation and acts by itself as a salt agent.

(2) Ion-exchante technique : This technique is used to separate the actinide ions. However, this is best suited for separation of the trans-americium elements and also for some of the preceding elements, especially for small amounts of material.

This method involves the following two steps :

(i) **Lanthanide-actinide separation :** The actinides as a group can be separated from lanthanides by using a cation-exchange resin. Strong HCl finds use as the eluting agent. The actinide ions tend to form chloride complexes more readily and hence get eluted first.

Separation of actinides from lanthanides is now carried out on an anion-exchange resin by using 10 MLiCI as eluant at elevated temperatures up to ~90°. With the exceptions of Gd, Ho and of Cm, the elution sequences follow the order of increasing atomic number. Hence, La gets absorbed least strongly.

(ii) **Separation of individual actinide elements :** In general the actinide ions can be separated from each other by removing from the cation exchange resin by carrying out elution with ammonium citrate, lactate, a-hydroxy isobutyrate and ethylene diamine tetra-acetate.

When the activity is made to plot against the number of drops of eluant, elution curves will be obtained as depicted in Fig. 4.5 Lw (atomic number, Z = 103) is expected to leave the column first, to be followed by No(Z = 102), and so on down the scale of atomic numbers. Elution positions for Md (Z – 101), Fm (Z = 100), and down to Am (Z = 95) have been depicted in the typical elution curves. Elution curves for lanthanides have been depicted in the same figure for comparison. A very striking similarity can be observed in the spacings of the corresponding elements in the two series (*e.g.*, Am and Eu, Cm and Gd, Bk and Tb etc.).

This similarity has helped scientists to predict the elution positions of the elements from Bk to Md before their discovery, and which also makes possible today to predict the elution positions of elements with Z = 102, 103, 104, 105 and 106.

COLOURS AND ABSORPTION SPECTRA OF ACTINIDE IONS

Colours and the wavelengths of the more prominent light absorption bands of actinide ions such as An^{3+}, AnO^{2+} and An_2^{2+} have been included in Table 4.5. From table it can be seen by comparing the colours of Ln^{3+} and An^{3+} ions, that *the pairs of ions which are having the same number of f-electrons (4f and 5f electrons) are having roughly comparable colours.* The pairs viz. (i) Nd^{3+} ($4f^3$-raddish) and U^{3+} ($5f^3$-raddish) and (ii) Gd^{3+} ($4f7$-colourless) and Cm^{3+} ($5f$-colourless demonstrates this point. Ce^{3+} ($4f^{\,1}$) and Pa^{4+} ($5f^1$) ions have been colourless.

The absorption spectra of actinide ions, in aqueous solution and in crystals, are having narrow bands in the visible, near ultraviolet and near infrared regions of the spectrum. The bands seen in case of actinide ions are ascribed to the electronic transitions between energy states within the $5f$-electron sub-shell. It can be seen, in general, that the absorption bands seen in case of actinide ions are about ten times sharper than those seen in case of lanthanide ions. The absorption spectra of lanthanide ions are because of the electronic transitions between energy states within the 4*f*-electron sub-shell.

Light absorption spectra of lanthanides are much more affected by environment. Charge-transfer phenomena take place more frequently and the bands are somewhat more intense in the actinides because of lower energy involved in transitions. Also, the overlap of 5*f*-orbitals with the ligand orbitals gets increased the intensity of absorption bands.

Magnetic Properties

In the 5*f*-series, Pu^{3+} and Am^{3+} ions have been found to exhibit the analogous behaviour as that for Sn^{3+} and Eu^{3+} ions in 4*f*-series.

It is more difficult to explain the magnetic properties of the actinide ions considerably than those of the lanthanide ions. The values of magnetic moments found experimentally have been found to be lower than those calculated by employing Russell-Saunders coupling *f*-scheme. This is probably because of the inadequacy of more subtle ligand field effects which are involved in bonding in the lanthanide complexes.

The equation used for the calculation of molar susceptibility, χ_M, is given as follows :

$$\chi_M = \frac{Ng^2\beta^2(J+1)}{3kT} + N\alpha$$

where N = Avogadro's number,

g = Lande splitting factor which is given by :

$$g = 1 + \frac{S(S+1) + J(j+1)\text{-}L(L+1)}{2J(+1)}$$

β = Bohr magneton= $\frac{eh}{2\pi mc}$ = 9.27 × 10–21 erg/gauss,

J = Total angular momentum of atom = | L + S |,

K = Boltzmann constant,

T = Absolute temperature,

a = Small, temperature independent term due to second order Zeeman effect.

Complex Formation

The degree of complex formation has been found to decrease in the following order :

$$An^{4+} > AnO_2^{2+} > An^{3+} > AnO_2^{+}$$

The complexing power of different singly charged and doubly chàrged anions has been found to follows the order given below :

Singly-charged anions : $F^- > NO^-_3 > Cl^-$

Doubly-charged anions : $CO^{2-}_3 > Ox^{2-} > SO^{2-}_4$

Actinides are also known to form complexes with a large number of organic substances.

COMPOUNDS

Thorium oxide or Thoria, ThO_2

Preparation

1. By heating hydroxide or nitrate,

$$Th(OH)_4 \rightarrow ThO_2 + 2H_2O$$

$$Th(NO_3)_4 \rightarrow ThO_2 + 4NO_2 + O_2$$

2. Crystalline form can be obtained on fusion with Na_2CO_3,

$$ThF_4 + 2Na_2CO_3 \rightarrow ThO_2 + 2CO_2 + 4NaF$$

Properties

1. It is a white powder.
2. It is hygroscopic and absorbs gases.
3. On heating it emits a pale blue light which becomes brilliant if mixed with 1% ceria (CeO_2).
4. It melts at 300°C and boils at 4400°C.
5. It does not dissolve in HCl or HNO_3 but in hot and conc. H_2SO_4 dissolves slowly, giving sulphate.

$$Th + 2H_2SO_4 \rightarrow Th(SO_4)_2 + 2H_2$$

6. On heating to a very high temperature, it reacts with boron and carbon to form boride and carbide.

Uses

1. **Incandescent gas mantle :** A gas mantle consists of a mesh-like structure of thorium oxide containing 1 per cent cerium oxide.
2. **Electric light filament :** Another important use of thoria is as an additive to tungsten filament.

Other Uses

Thorium oxide is also employed :

(i) as a starting material in the metallurgy of thorium,

(ii) as a component of refractories,

(iii) as a dehydrating catalyst in certain organic, and

(iv) as an optical glass polish.

Thorium Nitrate, $Th(NO_3)_4 . 12H_2O$

Preparation : It is obtained by dissolving thorium hydroxide in nitric acid and drying over sulphuric acid.

Properties

1. It is white deliquescent solid.
2. It is soluble in water and alcohol.
3. Its aqueous solution undergoes hydrolysis.

$$Th(NO_3)_4 + 4H_2O\ 4HNO_3 + Th(OH)_4$$

4. When heated it furnishes dioxide.
5. It forms double salts with ammonium and alkali nitrates.

Uses

1. In gas mantle industry.
2. In flash powder.
3. In the titration of the fluorides at 90°C using methyl red as indicator.

PLUTONIUM

History

The element with atomic number 94 is the fifth member of actinide series. The isotope plutonium-238 was identified by Seaborg *et al.* In 1941, it was synthesised by deuteron bombardment of uranium in 60 inch cyclotron. Plutonium exists in the form of natural and artificial isotopes. The isotope of great importance is plutonium –239 with half life period of 24, 360 years. It is produced in extensive quantities in nuclear reactors from natural uranium.

$$_{92}U^{238} + {}_{1}H^{2} \rightarrow {}_{93}NP^{238} + 2{}_{0}n^{1}$$

$$_{93}Np^{238} \xrightarrow[2d]{-\beta} {}_{94}Pu^{238} + {}_{-1}e^{0}$$

Occurrence

Plutonium is one of the transuranium elements which do not occur in nature in any appreciable quantities. Plutonium occurs along with neptunium in traces in the uranium ores, they are probably formed in the ores by means of neutron capture. The extraction of plutonium from uranium ores is not possible because of the negligible quantities.

Astrophysicists have suggested that probably the transuranium the transuranium elements exist in stellar systems. According to them, the synthesis of these elements on earth is a re-creation of the cosmic processes giving rise to all the elements.

The first transuranic element synthesised is neptunium (At. No. 93). In 1940 *McMillan, Kennedy, whal and seaborg* continued the search for the next transuranium element. They bombarded uranium –238 with deuterons from the 60-inch cyclotron at Berkley and were able to obtain anew isotope of neptunium, Np^{238} which decayed to plutonium –238.

$$_{92}U^{238} + {}_{1}H^{2} \rightarrow {}_{93}Np^{238} + 2{}_{0}n^{1}$$

$$_{93}Np^{238} \xrightarrow{\beta} {}_{94}Pu^{238}$$

In 1942 the element was named plutonium after the planet Pluto which is the second planet beyond Uranus. In 1941 Seaborg succeeded in identifuying plutonium –239 and confirmed its siginificance as a source of nuclear energy.

Uranium –238 target was bombarded with neutrons from the 60-inch cyclotron and uranium -238 was first converted into uranium –239 isotope which underwent decay forming neptunium–239 which is turn β decayed to form plutonium –239.

$$_{92}U^{238} + {}_{0}n^{1} \rightarrow {}_{92}U^{239} + \gamma$$

$$_{92}U^{239} \xrightarrow{\beta} {}_{93}Np^{239} \xrightarrow{\beta} {}_{94}Pu^{239}$$

This synthesis of plutonium is a turning point in the history of atomic energy as plutonium 239 was found to be a more effective nuclear fuel then uranium –235.

Preparation : (i) A sample of uranyl nitrate, $UO_2(NO_3)_2$ was embedded inside a large paraffin block to slow down the speed of neutron. A beryllium target which was exposed to full beam of deuterons from 60-inch cyclotron was placed in front of the paraffin block side which emitted neutrons.

$$_4Be^8 + {}_1H^2 \rightarrow {}_5B^9 + {}_0n^1$$

It is quite a difficult job to separate plutonium from such a complex mixture. However, Seaborg and his colleagues attempted the separation of plutonium by new techniques such as solvent extraction, bismuth phosphate coprecipitation and ion-exchange methods.

Separation of plutonium from this mixture is based on the comparative stability of the oxidation states of uranium and transuranium ions. The stability of these ions decreases in the order or

$$UO_2^{2+} = NPO_2^{2+} > PuO_2^{2+} > AmO_2^{2+}$$

$$Am^{2+} > Pu^{3+} >> Np^{2+} = U^{4+}$$

Since these elements differ in the stability of their various oxidation states they can be separated with the help of suitable oxidising or reducing agents followed by precipitation or solvent extraction method. Plutonium can be readily oxidised from +4 to +6 oxidation state using sodium bismuthate, potassium permanganate or dichromate. Plutonium can be reduced back to the +4 oxidation state by reducing agents such as ferrous salt as ferrous salt or nitrite.

(ii) The spent uranium fuel rod from the nuclear reactor is an important source of plutonium. The fuel rod with its protective aluminium jacket is dissolved in nitric acid. The acid solution contains uranyl ion, UO_2^{2+}, Pu^{4+} and other fission products which include various lanthanides and actinides. Plutonium and uranium are separated by ion-exchange method.

1. Tributyl Phosphate (TBP) Method

From the aqueous nitrate solution, MO_2^{2+}ions can be extracted into organic solvents. Tributyl phosphate (TBT) in kerosene can extract the M^{4+} ions from 6N nitric acid solution while M^{3+} ions from stronger nitric acid (10-16 N).

The order of extractability of the ions of these elements from 6N nitric acid solution decreases in the order of $Pu^{4+} > PUO_2^{2+}$: $Np^{4+} = NpO_2^{2+}$ >> Pu^{2+}; UO_2^{2+} >NpO_2^{2+} > PuO_2^{2+}.

The extraction of M^{3+} from 6N acid solution is very poor but it appreciably increases from 12N HCl or 16N HNO_3 solution in the order of Np < Pu < An < Cm < Bk.

The separation of plutonium from a mixture of uranium and plutonium is carried out, normally, by reducing plutonium to +6 to +4 oxidation state by a nitrite in the original solution. Pu^{4+} is comparatively more easily extracted into an organic phase than Pu^{6+}. Plutonium is transferred form organic phase to aqueous phase by further reduction of Pu^{4+} to Pu^{3+} by reducing agents such as hydroxylamine or sulphur dioxide.

(ii) Hexone (methyl isobutyl ketone) Method : Both uranium and plutonium in the +6 oxidation state are extracted by hexone (methyl isobutyl ketone) from the nitrate solution, while the other fission products are not extracted. The organic layer is washed with aqueous sulphur dioxide to reduce Pu^{6+} to Pu^{4+} which is stripped into the aqueous layer, while U^{6+} is left behind in the organic phase.

2. Coprecipitation Method

Uranium and transuranium elements in +3 and +4 oxidation states can be precipitated as fluorides or phosphates as fluorides or phosphates from the acid solution. They do not precipitate in their higher oxidation state of +6. The precipitation in +3 and +4 oxidation states can be prevented by complex formation with sulphate or other ions. Plutonium was initially separated by coprecipitation method. Bismuth phosphate is one of the coprecipitants which carries quantitatively Pu^{4+} as phosphate precipitate from moderately concentrated sulphuric or nitric acid solution. Under these conditions, uranium phosphate remains in solution, thus facilitating the separation of uranium from plutonium. The other fission products also mostly remain in solution.

Bismuth phosphate-Lanthanum fluoride cycle : The nuclear fuel solution consisting of uranium (UO_2^{2+}) plutonium (+6) and the fission products is treated with a nitrite for preferential reduction of Pu^{6+} to Pu^{4+}. From this solution uranium is separated by solvent extraction with hexone. From the aqueous phase, Pu^{4+} is coprecipitated with bismuth phosphate. $BiPO_4$. The precipitate is dissolved in nitric acid and re-precipitated several times. Finally the precipitate is dissolved in nitric acid and treated with sodium bismuthate, $NaBiO_3$ to oxidise Pu^{4+} to Pu^{6+}. Plutonium is then reduced from +6 to +4 state by ferrous salt. In this solution a lanthanum salt is added. Plutonium (+4) and lanthanum

(+3) are then coprecipitated as fluorides by treating with hydrofluoric acid. This is filtered and the precipitate is digested with sodium hydroxide and then dissolved in nitric acid. Pu^{4+} is oxidised to Pu^{6+} by a bromate, BrO_3^- or a persulphate, $S_2O_8^{2-}$. The resulting solutions is treated with hydrofluoric acid when lanthanum fluoride, LaF_3 alone precipitates. This is filtered and the filtrate is treated with sulphur dioxide to reduce Pu^{6+} to Pu^{4+}. In this solution lanthanum salt is dissolved and the lanthanum fluoride cycle is repeated.

3. Ion-Exchange Method

The spent uranium fuel rod from the nuclear reactor along with its protective aluminium jacket is as dissolved in nitric acid. The acid solution contains uranyl ion, UO_2^{2+}, Pu4+ nd other fission products which include various lanthanides and actinides. The separation of plutonium and uranium separately from this solution by ion-exchange method involves the following steps :

(i) **Lanthanide-Actinide separation :** The solution is mixed with a cation exchange resin such as Dowex -50 when the lanthanide and actinide cations are transferred to the resin phase. This mixture is then poured down a column of the same resin packed in a vertical glass tube. Strong hydrochloric acid is used as the eluting agent. The actinide ions readily form chloro complex anions and are desorbed first from the cation exchanger effecting the separation of actinides as a group from lanthanides.

(ii) **Separation of Uranium :** The eluate from the ion-exchange separation of lanthanides contains actinides as their chloro complex anions. This solution contains major quantity or uranium. The eluate is suitably treated to decompose the chloro complex anion and convert the actinides into their cations. The solution is treated with sulphuric acid when uranium alone forms a complex uranyl sulphate anion $[UO_2(SO_4)_3]^{4-}$. This solution is slowly poured down a column of long chained polystyrene anion exchange. The complex uranyl sulphate anions get adsorbed on the resin and the solution flowing out contains cations of the other actinides.

$$[UO_2(SO_4)_3]^{4-}(aq) + 4RNO_3(r) \rightleftarrows R_4[UO_2(SO_4)_3](r) + 4NO_3^-(aq)$$

Uranium is eluted by the use of a mixture of sodium chloride and hydrochloric acid or ammonium nitrate and nitric acid.

(iii) Separation of Individual Actinides : The solution obtained after the separation of uranium contains the remaining actinides which includes plutonium as their cations. The solution is poured down a cation-exchanger. In the case of actinides like lanthanides with increase in the atomic number the ionic radii decrease (due to actinide contraction) while the hydrated ionic radii increase. Larger the radius of the hydrated ion less strongly the hydrated ion is bound to the cation exchanger. Hence, the adsorption of actinides on the cation exchanger is in the order of their atomic number. The actinide ions are eluted using ammonium citrate, lactate or ethylene diamine tetra-acetate. The elution of actinides is in the reverse order of their atomic number. Therefore one of the separately collected eluate fraction contains plutonium from which it is precipitated as fluoride.

Metallurgy : Plutonium tetrafluoride is reduced with vapours of barium at about 1200°C to get the metal.

$$PuF_4 + 2Ba \rightarrow Pu + 2BaF_2$$

Plutonium-238 : When U-238 is bombarded with deuterons from a cyclotron, neptunium -238 is produced by (d, 2n) reaction.

Neptunium-238 is a β–emitter wiht a half- life of 2.10 days and yields plutonium-238 which is an a–emitter with a half period of 90 years. It was detected by *Seaborg et al.* in 1940. The reactions taking place may be shown as follows :

$$_{92}U^{238} + {}_{1}D^{2} \rightarrow {}_{93}Np^{238} + 2{}_{0}n^{1}$$

$$9_{93}Np^{238} \rightarrow {}_{94}Np^{238} + -{}_{1}e^{0}$$

Plutonium-241 : The isotope plutonium-241 is a valuable in sense that it is the parent of the most important isotope of the element americium. It is produced when U-238 is bombarded by alpha particles.

$$_{92}U^{238} + {}_{2}He^{4} \rightarrow {}_{94}Pu^{241} + {}_{0}n^{1}$$

It is a β–emitter with half-life period of 13 years, and gives rise to americium-241.

$$_{94}Np^{241} \rightarrow {}_{94}Am^{242} + {}_{-1}e^{0}$$

Metallic Plutonium : The metal may be obtained by the usual method of heating the fluoride in barium vapour. A better method due to *Morgan et al.* consists in a thermite type reduction of a mixture of PuF_4 and iodine by means of metallic calcium.

Properties

Plutonium is a silver-white metal (m.p.639.5,b.p.3235 °C.) Plutonium metal is unique as it possesses six allotropic forms, below its melting point. It also shows four ionic valence states in aqueous solutions. Pu^{+3} (blue), Pu^{+4} (yellow brown), PuO_2^{+2} (pink orange) and $PuO^+{}_2$ (colourless). The ion PuO_2^{+2} is unstable in aqueous solution. Plutonium is fairly strongly electropositive. Its standard oxidation potential is 1.88 volt ($Pu \rightleftarrows Pu^{3+} + 3e$).

Plutonium dissolves in hydrochloric acid to form plutonium (III) chloride, but is insoluble in HNO_3. When heated in air, plutonium forms the dioxide PuO_2, which is also produced when any of the plutonium compounds is ignited in air.

When treated with gaseous hydrogen fluoride, it gives a mixture of the trifluoride (purple) and trifluoride, (brown). The tetrafluoride and also the trifluoride, when heated with fluorine at 750°C, give the hexafluoride which may be condensed out of the vapour phase on a cold surface.

Chemistry of Plutonium in Aqueous Solution

The chemistry of plutonium in aqueous solution is of particular interest because of role it plays in the recovery of plutonium from irradiated uranium. Plutonium resembles uranium and neptunium in forming as many as four cations with oxidation states +3, +4, +5 and +6. The colours of these ions are given below :

Table 1 : Colours of plutonium cations

Cation	*Colour*
Pu^{3+}	Deep blue
Pu^{4+}	Yellow to brown
PuO_2^+	Colourless
PuO_2^{2+}	Pink-orange

The oxidation potentials, in volts, of the various plutonium ions in 1M strong acid, are shown below :

$$Pu^{3+} \xrightarrow{-0.97} Pu^{4+} \xrightarrow{-1.17} \underset{-1.04}{PuO^{+2}} \xrightarrow{-0.91} PuO_2^{2+}$$

We see that the oxidation potentials are all very nearly the same. This makes the oxidation reduction behaviour of plutonium in aqueous solution a complicated matter. The stability range of the trivalent plutonium is much greater than U^{3+} and Np^{3+} ions. Even such mild reducing agents as the ferrous ion and iodide ion can easily bring about the reduction of the higher states to Pu^{3+}. Plutonium (V) is the least stable of the four, and easily disproportionate to +4 and +6 states. Pu (IV) is also not stable and disproportionate to some extent. Hence, solutions containing other than +3 state or pure +6 state decompose spontaneously to give mixtures containing all the four ions in equilibrium with one another.

Hazards of Plutonium : (i) Working with plutonium is extremely hazardous for two different reasons : of plutonium enters person's body, the damage, near its points of deposition, is very serious because of irradiating the bone marrow and thus destroying the bloods cells. (ii) The second hazard with plutonium refers to its being a highly fissile material. Enriched plutonium could achieve criticality during the processing of used up nuclear fuels. Adequate precautions must therefore be take to prevent criticality.

Uses

Among the transuranic elements only plutonium has unique applications. Plutonium-239 possesses a dominant importance among the transuranic elements because of its use as an explosive ingredient in nuclear weapons. It also holds a key position in the development of development of industrial uses of nuclear energy. Pu-239 has the nuclear property of being fissionable with slow neutrons like U^{235}.

$${}_{94}Pu^{239} + {}_{0}n^{1} \rightarrow \text{fission products} + \text{Energy} + 2 \text{ or } 3 \text{ neutrons.}$$

It is also employed for the production of radioactive isotopes and higher transuranic elements like Am, Cm, etc.

URANIUM

General Remarks

Uranium is one of naturally occurring actinides. Among the actinides the chemistry of uranium is well understood, +6, +5, +4, +3 are the valencies which are known in the case of this metal. Among all these the hexavalent state is the prominent valence– so much so that prior to the knowledge of the actinides as a family, uranium was used to be included in group VI A of the periodic table.

In the + 6 state compounds tend to be covalent. This is shown by the volatility of UF_6. In this high valence state the element has (like Mo and W) a flair of forming large number of dioxouranium (UO_2^{2+}) salts and complexes.

In the +4 state, $U(OH)_4$ has practically no acid properties. Thus it reacts with acids to give uranium (IV) salts as UX_4. The uranium (IV) halides are crystalline solids and possess much higher melting points compared to those of uranium (VI). We know that in case of V, Mo and W with decrease in valence the tendency of forming oxo salts and oxo complexes decreases. This is true with uranium (IV) also. All its complexes are of the non-oxo types, e.g.

$$M_2(UCl_6], [NH_4]_4 [U(OX)_4] \text{ etc.}$$

Uranium (V) is very unstable and has a strong tendency to disproportionate.

$$2U(V) \rightarrow U(VI) + U(IV)$$

It forms oxo and also non-oxo compounds.

URANIUM DISCOVERY

Uranium was discovered by Klaproth in 1789 in the mineral Pitchblende, and was given this name by him after the planet Uranous. The discovery of uranium as a radioactive element was made in 1896 by Henry Becquerel.

Occurrence

1. Uranium does not occur free in nature. It is found in combined state in a few scarce minerals. Its proportion in the earth's crust is very small *i.e.* 0.000003%. Its important mineral is pitchblende in which uranium is present as U_2O_8. It is invariably contaminated with oxides of other metals such as iron, lead, calcium, radium, barium, arsenic, sulphur, silicon and number of other metals like thorium, vanadium and rare earths. It occurs much in Congo, North West Canada and United States.
2. Carnotite is potassium uranyl orthovanadate, $K(UO_2)VO_4.3H_2O$: it occurs in Colorado, U.S.A.
3. Autunite is potassium uranyl orthophosphate, $K(UO_2)PO_4.8H_2O$; it is found in Congo.

In India the principal source of uranium ore is the Jadugada mine in Bihar. In 1964, rich deposits of high grade uranium ore have been found in Bhatin area of Bihar. Another important source of uranium in India has been the monazite sands of Travancore.

Extraction

From Pitchblende

Pitchblende is first of all concentrated by gravity process for removing sand, clay etc. It is then roasted in excess of air whereby S, As and Sb are removed as their volatile oxides. If silver is present in the mineral, it is removed as insoluble AgCl by roasting the mineral with NaCl at about 800°C. The roasted ore is now made to digest either with sulphuric acid in presence of an oxidising agent like conc. HNO_3 or MnO_2 (acid digestion process) or with alkali ($NaHCO_3 + Na_2CO_3$) (alkali digestion process) for extraction of uranium.

(a) Acid digestion process : When the roasted ore is made to digest with dil. H_2SO_4 in the presence of MnO_2 or concentrated nitric acid (an oxidising agent) for about 24 hours, barium, lead and radium are removed as their insoluble sulphates whereas uranium forms soluble uranyl sulphate, UO_2SO_4.

$$2U_3O_8 + 6H_2SO_4 + O2 \rightarrow \underset{\text{Uranyl sulphate}}{6UO_2SO_4} + 6H_2O$$

When the filtrate having uranyl sulphate is treated with an excess of sodium carbonate, uranyl sulphate gets converted into soluble uranyl carbonate, Na4UO2$(CO_3)_3$ whereas other metals such as iron, aluminium, cobalt, manganese, lead etc. get precipitated.

$$UO_2SO_4 + 3Na_2CO_3 \rightarrow \underset{\text{Sod. uranyl carbonate}}{Na_4UO_2(CO_3)_3} + Na_2SO_4$$

The filtrate having sodium carbonate is acidified with hydrochloric acid so as to decompose it in to uranyl chloride, UO_2Cl_2. The resulting solution is saturated with H_2S gas so that lead, copper and arsenic get precipitated as their sulphides. The filtrate is made to boil to remove hydrogen sulphide and finally the uranium gets precipitated as ammonium diuranate. $(NH_4)_2U_2O_7$ by adding ammonia. The ammonium diuranate is ignited in air to yield U_3O_8.

$$NaUO_2(CO_3)_3 + 6HCl \rightarrow \underset{\text{Uranyl chloride}}{UO_2Cl_2} + 4NCl + 3H_2O + 3CO_2$$

$$2UO_2Cl_2 + 6NH_4OH \rightarrow \underset{\text{lignite}}{(NH_4)_2U_2O_7} + 4NH_4Cl + 3H_2O$$

$$3(NH_4)U_2O_7 \rightarrow 2U_3O_8 + 6NH_3 + 3H_2O + O_2$$

(b) Alkali digestion process : When the pasted ore is digested with $Na_2CO_3 - NaHCO_3$ mixture solution, U_3O_8 present in the mineral gets converted into soluble sodium uranyl carbonate. The resulting solution is now made to treat with sodium hydroxide which would precipitate uranium as sodium diuarante, $Na_2U_2O_7$.

$$2U_3O_8 + 6Na_2CO_3 + 12NaHCO_3 + O_2 \rightarrow 6Na_4UO_2(CO_3)_3 + 6H_2O$$

$$2Na_24UO_2(CO_3)_3 + 6NaOH \rightarrow \underset{\text{sod. diuranate}}{Na_2U_2O_7} \downarrow + 6Na_2CO_3 + 3H_2O$$

The precipitate of sodium diuranate is first of all dissolve in dil. H_2SO_4 and the solution is then treated with ammonium hydroxide to obtain the precipitate of ammonium diuranate. Precipitate of ammonium diuranate when ignited in air, yields U_3O_8.

$$Na_2U_2O_7 + H_2SO_4 \rightarrow Na_2SO_4 + H_2U_2O_7$$

$$H_2U_2O_7 + 2NH_4OH \rightarrow \underset{\text{Amm. diuranate}}{(NH_4)_2U_2O_7} \downarrow + 2H_2O$$

2. From Carnotite

The concentrated ore is heated with concentrated nitric acid at about 100°C and the solution is neutralised with caustic soda. The resulting solution is made of treat with barium chloride and sulphuric acid to remove radium as radium sulphate by co-precipitation with barium sulphate while uranium and vanadium get converted into soluble uranyl and vanadyl sulphates.

The filtrate having uranyl and vanadyl sulphates is treated with an excess of boiling sodium carbonate solution for precipitating iron, calcium and aluminium. During this treatment uranyl and vanadyl sulphates get converted into soluble sodium uranyl carbonate and sodium vanadate respectively.

$$UO_2SO_4 + Na_2CO_3 \rightarrow Na_4UO_2(CO_3)_3 + Na_2SO_4$$

$$2VOSO_4 + 5Na_2CO_3 + \frac{1}{2}O_2 \rightarrow 2Na_3VO_4 + 2Na_2SO_4 + 5CO_2$$

$$Na_4UO_2(CO_3)_3 + 6HNO_3 \rightarrow UO_2(NO_3)_2 + 4NaNO_3 + 3H_2O + 3CO_2$$

$$2UO_2(NO_3)_2 + 6NaOH \rightarrow Na_2U_2O_7 + 4NaNO_3 + 3H_2O$$

The precipitate of sodium uranate, so formed gets converted into ammonium uranate by ammonia (as in acid digestion process) and the ammonium uranate gets ignited to yield uranium oxide, U_3O_8.

Reduction of U_3O_8 : Uranium oxide and other uranium compounds can be reduced readily by the common reducing agents like carbon, aluminium, calcium, sodium, etc. However, the preparation of completely pure metal as needed in nuclear reactions is very difficult partly because of the strong tendency of uranium to form the carbide and partly due to its affinity for oxygen and nitrogen. However, the two successful methods for the preparation of completely pure uranium are outlined as follows :

(i) The metallic uranium which is used for nuclear reactors is generally prepared by the reduction of uranium tetrafluoride with metallic calcium or magnesium.

$$UF_4 + 2Ca \rightarrow U + 2CaF_2 + 562.4 \text{ kJ}$$

Uranium tetrafluoride required for the above step is prepared by treating uranium dioxide, UO_2 with gaseous hydrogen fluoride at 400-500°C, the UO_2 in turn is obtained by the reduction of U_3O_8 with hydrogen. Hence, U_3O_8 constitutes the starting compound for the production of pure uranium required in nuclear reactors. The whole series of reactions can be put as below,

$$U_3O_8 \xrightarrow{H_2} UO_2 \xrightarrow{HF} UF_4 \xrightarrow{Ca} U$$

Calcium needed for the process is prepared by the reduction of calcium oxide with aluminium and then sublimating the so formed metal in vacuum. In the process, uranium is obtained in the molten state which is the made to solidly in mould of sintered fluorite. The solidified uranium is again melted in vacuum in uranium crucible and therefore 99.98% pure uranium can be obtained.

(ii) *Alumino-thermic process :* It is possible to obtain 98-99% of pure uranium by reducing $U3O_8$ with aluminium.

$$3U_3O_8 + 16Al \rightarrow 8Al_2O_3 + 9U$$

(iii) *Reduction of oxide with carbon :* Moissan obtained uranium by heating the oxide, U_3O_8 with carbon in an electric furnace.

$$U_3O_8 + 8C \rightarrow 3U + 8CO$$

Reduction with magnesium or calcium can also be carried out. Uranium oxide is made of heat with magnesium or calcium when uranium of 98.99 percent purity is obtained. Burger was able to reduce the oxide by the vapour of calcium in vacuo.

(iv) *Peligot's method* : Peligot prepared the metal by reducing UC13 mixed with potassium chloride and sodium in a porcelain crucible. The temperature was raised rapidly and the cold product was lixiviated with water.

$$UCl_4 + 4Na \xrightarrow{KCl} U + 4NaCl$$

3. Electrolytic Process

The main method for preparing large amounts of uranium is by the electrolysis of fused salts, like those of tetravalent uranium mixed with alkali halides and alkaline earth halides at fairly low temperatures. Hence, the electrolyte consists of KUF_5 mixed with NaCl and $CaCl_2$. The eletrolysis is done in a graphite crucible which acts as the anode and the cathode is made of molybdenum strip and gets suspended in the centre of crucible. The electrolysis is done at 775°C when uranium is collected on the cathode.

Properties

Uranium is bright white lustrous metal. It slowly tarnishes in air. The density of pure uranium is 19.03 and its melting point is 3500°C. Pure uranium is ductile and malleable and can be moulded at room temperature by hammering and rolling. On heating, it first becomes brittle and then elastic at higher temperature; this change during heating occurs due to the allotropic transformation.

$$\alpha - U \xrightarrow{600^\circ} \beta - U \xrightarrow{700^\circ} \gamma - U$$

Uranium is a radioactive element and is having three isotopes.

U^{238}, U^{235} and U^{234}

Chemically, uranium is highly reactive. However, its important chemical reactions of uranium are given as follows :

(i) **Action of air :** In air at ordinary temperature its bright white surface is rapidly covered first with a yellow and then a black non-protective film; while on heating in air it burns to yield U_3O_8.

$$3U + 4O_2 \rightarrow U_3O_8$$

(ii) **Action of water :** The reaction of uranium with water is complex, uranium decomposes the boiling water to yield the dioxide.

$$U + 2H_2O \rightarrow UO_2 + 2H_2$$

Then hydrogen so formed reacts with uranium to form finely divided uranium hydride which brings about disintegration.

$$2U + 3H_2 \xrightarrow{250°-300°} 2UH_3$$

Uranium may react with water even at room temperature, when the metal is finely divided.

(iii) **Action of halogens :** Uranium reacts rapidly with all the halogens at different temperature to yield the corresponding tetrahalide.

$$U + 2F_2 \xrightarrow{\text{room temp.}} UF_4$$

$$U + 2Cl_2 \xrightarrow{180°} UCl_4$$

$$U + 2Br_2 \xrightarrow{240°} UBr_4$$

$$U + 2I_2 \xrightarrow{290°} UI_4$$

(iv) **Action of acids :** Uranium dissolves readily in dilute acids to yield uranium (IV) salts.

$$U + 4HCl \rightarrow UCl_4 + 2H_2$$

$$U + 2H_2SO_4 \rightarrow U(SO_4)_2 + 2H_2$$

With concentrated sulphuric acid, it evolves sulphur dioxide which with concentrated nitric acid evolves nitrogen oxides.

$$U + 2H_2SO_4 \rightarrow USO_4 + 2H_2O + SO_2$$

$$3U + 8HNO_3 \rightarrow U_3O_8 + 4NO + 4NO_2 + 4H_2O$$

(v) **Action of alkalies :** Uranium does not get attacked by alkalies.

(vi) **Action of metals and non-metals :** It reacts directly with most of the elements in the periodic table, especially at elevated temperatures, e.g.

$$U + 2Z \rightarrow UZ_2 \text{ (where, Z = C, S, Se or N)}$$

Uses

1. *As a source of nuclear energy in nuclear reactors.*
2. ***In ceramics :*** *Compounds of uranium are used for colouring glass and pottery, giving them yellow and orange glazes.*

3. ***For alloys :*** *Uranium alloys are pyrophoretic.*
4. ***Incandescent lamps :*** UO_2 *is used in incandescent lamps (used in picture projection and in photography).*
5. ***Textile printing :*** *Uranium salts are used as mordants for wood, silk and calico printing.*
6. ***As catalyst :*** *Uranium metal and uranium carbide are used as catalyst for the synthesis of ammonia by Haber's process.*

$$N2 + 3H2 \xrightarrow{\text{U Or UC2}} 2NH3$$

7. ***Photography :*** *Uranium acetate or nitrate is used for intensification of weak negatives.*
8. ***Rubber :*** *Addition of uranium borate improves the appearance, increases the tensite strength and resistance to oxidation to rubber.*
9. *Medicines : Uranium salts are found useful in medicines.*
10. ***Analytical chemistry :***

(i) *Uranous sulphate is used in the indirect titration of* HNO_3, H_2CrO_4 *etc.*

(ii) *Uranyl nitrate has been used as reagent for volumetric estimation Of phosphate and arsenate.*

(iii) *Uranyl acetates is employed for the detection of sodium.*

COMPOUND OF URANIUM

Uranium Hydride–UH_3

Preparation

Uranium combines with hydrogen at an elevated temperature 500-600°C and at a pressure of 100 atmospheres forming a hydride, UH_3.

$$2U + 3H_3 \xrightarrow[100\text{ atm}]{500-600°} 2UH_3$$

Properties

1. *It is a white brittle solid substances.*
2. *It is a good conductor of electricity like the uranium metal.*
3. *The hydride is very stable and does not decompose when heated under a high pressure even to a high temperature.*

4. *The X-rays study of the substance has revealed a peculiar hydrogen linkage :* U–H–U

Uses

1. *In the analysis of heavy water.*
2. *In the preparation of pure, finely divided dispersed medium of uranium.*

Uranium Tetrafluoride–UF_4

Preparation

1. *It is prepared by the action of HF on UO_2 or U_3O_8.*

$$UO_2 + 4HF \rightarrow UF_4 + 2H_2O$$

$$U_3O_8 + 16HF \rightarrow 3UF_4 + 6H_2O + 2f_2$$

2. *By the action of fluorine on uranium pentachloride.*

$$2UCl_5 + 5F_2 \rightarrow UF_4 + UF_6 + 5Cl_2$$

Properties

1. *It is a green, insoluble, amorphous powder.*
2. *It is stable to heat and melts at 1000°C.*
3. *It is insoluble in water.*
4. *It is sparingly soluble in other acids but readily soluble in nitric acid.*
5. *It changes to UO_2F_2 by moisture.*

Uranyl Nitrate, $UO_2(NO_3)_2.9H_2O$

Preparation

1. *By dissolving uranium oxide in nitric acid.*

$$UO_3 + 2HNO_3 \rightarrow UO_2(NO_3)_2 + H_2O$$

2. *By treating the solution of uranyl chloride with silver nitrate.*

$$UO_2Cl_2 + 2AgNO_3 \rightarrow UO_2(NO_3)_2 + 2AgCl$$

Properties

1. *it is soluble in water and ether. The aqueous solution undergoes hydrolysis on boiling.*
2. *It is deliquescent.*

3. *On heating with NH_4Cl it forms uranous oxide.*
4. *It forms a large number of double nitrates and complex nitrates with alkali nitrates, i.e. $NaUO_2(NO_3)_3$. Ammonium nitrate forms ammonium uranyltrinitrate with uranyl nitrate $NH_4UO_2(NO_3)_3$.*

Uses

1. *In photography.*
2. *As a laboratory reagent for the determination of phosphate.*
3. *For the detection of sodium with zinc or magnesium acetate.*

Uranium Hexafluoride-UF_6

It is one of the most important uranium compounds employed in nuclear technology. It has been most extensively used in the separation of uranium isotopes by the diffusion process because of its being most volatile.

Preparation

1. Uranium hexafluoride is generally prepared by the direct action of fluorine on UF_4.

$$UF_4 + F_2 \rightarrow UF_6$$

2. By reacting UF_4 with dry oxygen at about 800°C.

$$2UF_4 + O_2 \rightarrow UF_6 + UO_2F_2$$

Properties

The physical properties of uranium hexafluoride are remarkable. It is a colourless, snow-like volatile solid, which sublimes at 56.4°C.

Chemically, UF_6 is highly reactive and behaves like a powerful fluoridating agent. Examples are :

$$UF_6 + 2H_2O \rightarrow UO_2F_2 + 2HF$$

$$UF_6 + H_2 \xrightarrow{600°C} UF_4 + 2HF$$

Pentavalent Uranium : Uranium (V) has strong tendency to disproportionate 2U(V) → U(IV) + U(IV). It is also oxidised with great ease and rapidly to uranium (VI). The ground state electronic configuration of uranium (V) is $[Rn]5f^1$. Most of the uranium (V) compounds may be considered to contain covalently bound uranium (V), some being of the following oxo-types : UO_4^{-3} UO_3^-, UO_2^+ and UO_+^3. The uranium pentachloride is obtained by chlorination of UCl_4. Since of pentachloride easily disproportionates, it is recovered by cooling of the reaction product.

$$2UCl_4 + Cl_2 \xrightarrow{520-550°C} 2UCl_5$$

$$2UCl_5 \xrightarrow{Vacuum(120-150°C)} UCl_4 + UCl_6$$

A number of salts of the anion $[UCl_6]^-$ have been prepared by the oxidation of (UCl6)2- by chlorine in nitromethane. The salts isolated include $Cs[UCl_6]$, $Me_4N[UCl_6]$, $Ph_4N[UCl_6]$ etc. The largest group of uranium (V) compounds are the alkoxides and halide alkoxides :

$$UCl_4 + 5ROH \rightarrow U(OR)_4 \xrightarrow[NaOR]{O2} U(OR)_5$$

$$U(OEt)_4 + \frac{1}{2}Br2 \rightarrow U(OEt)4Br \xrightarrow{NaOEt} U(OEt)_5 + NaBr$$

$$UCl_5 + 2ROH + 5NH_3 \rightarrow U(OR)_5 + 5NH_4Cl$$

Uranium (V) oxotrichloride has been obtained as shown below :

$$UO_2Cl_2 + MoOCl_3 \rightarrow UOCl_3 + MoO_2Cl_2$$

Quadrivalent Uranium : On moving from +6 state to the +4 state the acidic properties have practically disappeared. The dioxide is prepared by the reduction of UO_3 or U_3O_8 with hydrogen of carbon. It is high melting (2176°C) brown solid.

The tetrahalides are generally made by the action of the halogen on uranium, or by the action of the halogens of a mixture of U_3O_8 and carbon. The halides are all high melting solids. UCl_4, 567°C : UI_4 500°C. In fused state these conduct electric current and these have good degree of ionic character.

Acid solution of uranyl salt can be reduced to green U^{4+} by $SnCl_2$ or zinc amalgam. But U^{4+} tends to be readily oxidised. The acetate and oxalate are reasonably stabel when dry. Uranium (IV) acetate is obtained as light green precipitate by the reduction of uranyl acetate by amalgam zinc in glacial acetic acid.

$$UO_2(CH_3COO)_2 + Zn + 4CH_3COOH \rightarrow Zn[U(CH_3COO)_6] + 2HO$$

$$Zn[U(CH_3COO)_6] + 2HCl \rightarrow U(CH_3COO)_4 + ZnCl_2 + 2CH_3COOH$$

The liberated water is removed by acetic anhydride.

Hexachloro complexes of uranium (IV) have been prepared by the reduction of uranyl chloride in presence of HCl and the complex isolated as pyridinium or dipyridyl cation : $(pyH)_2 [UCl_6]$, dipyridyl-$H_2[UCl_6]$ etc. These form green crystals and are paramagnetic as expected for a $5f_2$ system. The insoluble oxalate of uranium (IV), $U(OX)_2$ is very soluble in alkali Oxalate solution due to the formation of a soluble complex $K_4[U(OX)_4]$.

Trivalent Uranium : The +3 oxidation state in uranium is mostly known in trihalides. When UF_4 is heated in H_2, HF is given off and a red-brown solid UF_3 is obtained. UCl_3 is obtained by reducing the tetrachloride with hydrogen. It dissolves in water to purple-red liquid which evolves hydrogen, turns green and forms $U(OH)_4$. The tribromide has also been obtained by the action of hydrogen on UBr_4.

RHENIUM

Discovery

Rhenium element was discovered by *Naddack* and *Tacke* in 1925 after a planned and careful search in possible minerals. This element was suspected to be present only in traces in the minerals of neighbouring elements and therefore, concentrated to obtain the element in detectable quantities which were prepared by laborious methods. In 1928, these investigators were able to isolate one gram of pure rhenium from 660 kilogram of molybdenite mineral. Now this metal is available in commercial quantities.

Occurrence

There are no minerals of rhenium. Investigations have shown that silicates, niobates, tungstates and sulphide minerals usually contain trace amounts of the metal. The metal has been found to be associated with platinum ores. The ores containing rhenium are as follows :

Table 2

Ores	*Rhenium content in g/tonne*
Molybdenite	21.0
Osmiridium	1.0
Lead selenite	0.8
Gadolinite	1.1
Native platinum	0.8

Concentration of Rhenium

On roasting of molybdenite, the rhenium present as disulphide is converted into three oxides, Re_2O_7, ReO_3 and ReO_2 which being appreciable volatile go into the gas phase. Thus 75-95% of rhenium passes into the gas phase. Under favorable conditions, these oxides condense in the form of smoke particles and are separated by electrostatic filters or gas washing.

Dissolution of Rhenium from the Flue Dust

Following methods are employed to bring rhenium into solution phase from the flue dust, concentrated as given above.

(a) **Direct leaching with water :** Re_2O_7 the main rhenium constituent of the flue dust, being soluble in water, yields a solution of rhenium. The residue containing dioxide and trioxide is then treated with dil. HCl or H_2SO_4 to extract 30-60% of the remaining rhenium.

(b) **Leaching with water containing oxidising agent :** A suspension of flue dust in water, when reacted with an oxidising agent like chlorine, bromine or hypochllorite brings about 90% of rhenium into solution.

(c) **Leaching with after sintering :** The flue dust is sintered at about 600°C often being mixed with one and half times its weight of slaked lime. Calcium perrhenate, a readily soluble salt, is formed which is leached with warm water (70°C) and after two hours rhenium solution is filtered out.

Extraction of Rhenium form Solution

1. The rhenium solutions obtained from the above methods are mainly concentrated and then KCl is added. A crude potassium perrhenate ($KReO_4$) is precipitated. It is to, be recrystallized several times before it is obtained in the pure state.

   The crude potassium perrhenate is dissolved in the amount of boiling water containing H_2SO_4. The solutions filtered off to remove insoluble impurities. On cooling and adding more of KCl to the solution comparatively pure $KReO_4$ is obtained. After repeating the process several times very pure potassium perrhenate is obtained.

2. Molybdenite mineral contains two to four parts of rhenium in million parts of the ore. This is the most convenient source for the extraction of rhenium. The sulphide mineral is powdered and treated with concentrated HNO_3 at 70 to 80°C which converts molybdenum disulphide into molybdenum trioxide and rhenium to Re_2O_7.

   This solution is diluted with water and insoluble molybdenum trioxide MoO_3 is filtered out. The filtrate is then concentrated by evaporation and then treated with conc. H_2SO_4. The sulphuric acid solution is then distilled first in stream of CO_2 up to 170°C and finally in hydrogen chloride gas between 170-200°C. The distillate contains rhenium heptoxide Re_2O_7 with small amounts of molybdenum trioxide. From this distillate rhenium is separated in the following way :

   The distillate containing the two oxides is treated with the solution of oxine or 2, 2 dipyridyl so that molybdenum is completely precipitated as the complex salts. The filtrate is then treated with the solution of nitron acetate so that nitron perrhenate is pecipitated. This is filtered out, dried again, and ignited to obtain rhenium heptoxide.

Preparation of Metal

1. **Hydrogen reduction of potassium perrhenare :** Rhenium metal is usually obtained by reducing $KReO_4$ by hydrogen above is melting point (550°C).

   $$2KReO_3 + 7H_2O \rightarrow 2Re + 5KOH + 6H_2O$$

   Sometimes, H_2 at the pressure of 20-100 atmospheres is employed for this purpose. A relatively low temperature of 200-300°C is required .

2. **Electrodeposition :** On electrolysing a solution of potassium perrhenate in H_2SO_4 rhenium is deposited at the cathode. Anode is made up of platinum while cathode is of tantalum or graphite. Metal of 99% purity is obtained by this method.

3. **High purity :** Rhenuim metal is made by the hydrolysis of $ReCl_5$ in aqueous medium to ReO_2 and subsequent reduction of the product with H_2.

PROPERTIES OF RHENIUM

Physical

Finely divided rhenium is a grey-black powder but the compact metal is white like platinum. Pure rhenium is quite soft and malleable in the compact state.

Rhenium melts at 3180°C and boils at about 5530°C. The metal crystallises with hexagonal close packed structure.

Chemical

(a) **Action of oxygen :** Finely powdered metal oxidises easily in air at 150°C. The compact metal is oxidised by oxygen when heated above 1000°C.

$$4Re + 7O_2 \rightarrow 2Re_2O_7$$

(b) **Action of halogen :** Rhenium metal is directly attacked by halogens to form halides. Fluorine reacts at 12°C to form the hexafluoride ReF_6, chlorine reacts with rhenium at high temperature to form pentachloride $ReCl_5$ and trichloride $ReCl_3$ while bromine reacts with rhenium at 500°C to form tribromide $ReBr_3$. A stable iodide of rhenium has been obtained in pure state.

$$Re + 3F_2 \xrightarrow{125°C} ReF_6$$

$$2Re + 4Cl_2 \xrightarrow{500°C} ReCl_5 + ReCl_3$$

$$2Re + 3Br_2 \xrightarrow{500°C} 2ReBr_3$$

(c) **Action with acids :** Rhenium is slowly attacked by concentrated H_2SO_4 and the action is fast with HNO_3 and converts rhenium to perrhenic acid.

$$2Re + 14HNO_3 \rightarrow 2HReO_4 + 14NO_2 + 6H_2O$$

$$2Re + H_2SO_4 \rightarrow 2HReO_4 + 7SO_2 + 6H_2O$$

(d) **Action with non-metals :** Rhenium combines with sulphur to form the disulphide ReS_2. With phosphorus it forms a number of phosphides Re_2P, ReP, ReP_2, Rep_3, ; with arsenic it forms arsenide Re_3As_7 while with silicon the silicide $ReSi_2$.

(e) **Action with metals :** Rhenuim readily forms alloys with Nb, Ta, Mo, W, Fe, Co, Ni and platinum metals. Rhenium with tungsten appears to form the compound Re_3W_2.

(f) **Action with fused alkalies :** When rhenium is fused with alkali metal hydroxides in presence of oxygen, alkali metal perrhenates are obtained.

$$Re + 4KOH + 7O_2 \rightarrow 2HReO_4 + 2H_2O$$

Compounds of Rhenium

Compounds of heptavalent rhenium : When powdered rhenium is gently heated in oxygen it forms the oxide Re_2O_7 ; the vapours on cooling form yellow crystals. It is very soluble in water and the dilute solutions contain the acid known as perrhenic acid $HReO_4$.

$$Re_2O_7 + H_2O \rightarrow 2HPeO_4$$

Perrhenic acid : Perrhenic acid $HReO_4$ is only known in dilute solutions which on evaporation forms the oxide Re_2O_7. Perrhenic acid is a monobasic acid. It is a strong acid and forms colourless salts. These salts do not have water of crystallization.

Compounds of Hexavalent Rhenium

Rhenium trioxide : Rhenium trioxide is prepared by heating $Re2O_7$ with rhenium metal in sealed tube at 300°C. It is a red solid which forms a cubic lattice.

Compounds of Pentavalent Rhenium

Rhenium pentachloride : $ReCl_5$ is the only salt of pentavalent rhenium. It is formed along with $ReCl_5$ when rhenium is heated in chlorine. It is separated from the trichloride by sublimation in high vacuum. $ReCl_5$ is a brownish black solid which loses chlorine when it is heated in an atmosphere of nitrogen.

$$2ReCl_5 \rightarrow 2ReCl_3 + 2Cl_2$$

On being heated in oxygen, it forms the oxychloride $ReOCl_4$ and ReO_3Cl.

Rhenium pentachloride dissolves in conc. HCl to form hexachlororhenic acid.

$$2ReCl_5 + 4HCl \rightarrow 2H_2ReCl_6 + Cl_2$$

Compounds of Tetravalent Rhenium

Rhenium dioxide : ReO_2 is prepared by the reduction of Re_2O_7 with hydrogen at 300°C.

$$ReO_7 + 3H_2 \rightarrow 2ReO_3 + 3H_2O$$

Rhenium dioxide is brownish black solid which burns in oxygen to form the heptaoxide Re_2O_7.

$$4ReO_2 + 3O_2 \rightarrow 2Re_2O_7$$

When dioxide is heated in H_2 at 800°C, it is reduced to the metal.

$$ReO_2 + 3H_2 \xrightarrow{800°C} Re + 2H_2O$$

When rhenium dioxide is heated in vacuum, it decomposes to rhenium metal and Re_2O_7.

$$7ReO_2 \rightarrow 3Re + 2Re_2O_7$$

PALLADIUM

History

Palladium was discovered by *Wollaston* in 1803 in the crude platinum metal. The name palladium was given to the element in honour of the recently discovered asteroid "*Pallas*".

Occurrence

It is a rare metal. It is found to occur in the free state in the form of octahedral cubic crystal in Brazil and Ural mountain. Seleno palladium is another native source of the element. In the combined state it is found to occur as :

1. *Braggite* (Pt, Pd, Ni)S
2. *Stibiopalladinite,* Pd_3Sb
3. *Potarite*, PdHg
4. *Porpazite,* which is palladium gold ore.

Traces of palladium are found in various platinum ores. In addition to these it is also present is the earth's crust.

Extraction

1. From Platinum Ores

The palladiferrous material is treated with aqua regia. The insoluble residue is separated by filtration. The acid filtrate is trreated with an excess of NH_4Cl and ethanol. Platinum and iridium are precipitated as ammonium chloro-compounds by adding NH_4Cl. The palladous salts remains in solution. The precipitate is separated and from the filtrate platinum and iridium are completely removed by adding NH_4Cl and alcohol. The filtrate is treated with mercuric cyanide to precipitate the palladium as cyanide. Pure Pd is obtained by igniting the cyanide.

2. Wollaston's Process

Native platinum is treated with aqua regia and precipitated as $(NH_4)_2PtCl_6$ by adding NH_4Cl by adding. The filtrate is reduced with zinc or iron. This is treated with dil. HNO_3 and filtered. Copper, iron, lead and zinc go into KNO_3. Palladium and platinum go into solution. K_2PdCl_4 is crystallised out by evaporation of the solution. The crystals of potassium palladium chloride is treated with KOH. Palladium oxide is obtained which on ignition in a stream of hydrogen yields palladium.

Metallurgy

Palladium metal is obtained by ignition of :

(i) *Palladium cyanide*, $Pd(CN)_2$

$$Pd(CN)_2 \xrightarrow{\text{ignite}} Pd + (CN)_2 \uparrow$$

(ii) *Palladium iodide*,

$$PdI_2 \xrightarrow{\text{ignite}} I_2 \uparrow$$

(iii) *Ammonium salt*, (NH4)2PdCl4

$$(NH_4)_2PdCl_4 \rightarrow Pd + 2NH_3 + 4HCl$$

Properties

Physical

1. *Palladium stands between silver and platinum in appearance. Its density is 12.0.3.*
2. *It softens before melting (M.P. 1555°C).*
3. *Among the platinum metals it is most fusible.*
4. *It is malleable, ductile and can be welded at red heat.*
5. *Palladium has an adsorptive capacity for many gases especially for hydrogen. When finely divided it adsorbs 700 or more times its own volume of hydrogen. A part of adsorbed hydrogen is probably held in solid solution in the metal. The adsorbed hydrogen is very active and hence palladium has a pronounced catalytic activity.*

Palladium with adsorbed hydrogen is very efficient reducing agent.

Colloidal palladium adsorbs 3 to 8 times as much hydrogen as ordinary palladium.

Chemical

Palladium is chemically inert.

1. *It is not attacked by mineral acids except HNO_3 which does slowly dissolve it. In the finely divided state palladium is slowly attacked by HCl in presence of oxygen.*
2. *When heated to red heat in air, it is oxidised to palladium monoxide. Molten metal dissolves oxygen which is given off on solidification as in the case of silver.*
3. *At red heat fluorine and chlorine attack the metal and form difluoride and dichloride.*
4. *It shows valencies of 2, 3, 4 but bivalent compounds are very stable.*
5. *Palladium is rapidly attacked by fused alkali oxides and especially peroxides.*

Uses

Palladium stands next to platinum in applications.

1. *Colloidal palladium is used as catalyst in the hydrogenation of nitro-groups, aldehydes, ketones and many organic compounds. It also finds application in the cracking of petroleum.*
2. *It is also used in electroplating.*
3. *It is used in dental, jewellery and electrical industries as a substitute for platinum.*
4. *Palladium compounds are used as toning agents in photography.*
5. *Palladium nitrate is used in the identification and estimation of carbon monoxide in mines and garages.*

Compounds

Palladium shows valence states of 1,2, 3 and 4 in its compounds.

Halides

Chlorides : Three chlorides namely PdCl, $PdCl_2$ and $PdCl_3$ are known. The *monochloride*, (PdCl) is obtained by heating $PdCl_2$ at red heat. It is reddish-brown crystalline solid.

Palladous chloride : ($PdCl_2$) is made by the action of Cl_2 on the metal heated to red heat. It is hydroscopic and soluble in water. From the aqueous solution crystals of $PdCl_2.2H_2O$ are obtained. In the anhydrous conditions, the structure of the compound may be given as below. $PdCl_2$ is soluble in HCl acid or alkali metal chloride solution due to the formation $[PdCl_4]^{2-}$

Palladium trichloride : ($PdCl_3$) is unstable and is formed when palladium monoxide is treated with HCl.

Palladium difluoride : PdF_2 obtained by adding HF to a concentrated solution of $Pd(NO_3)_2$. It is a brown powder and is slightly soluble in water.

Palladium trifluoride : (PdF_3) is produced by the action of fluorine on $PdCl_3$ at 200-250°C,

Oxides

Three oxides of palladium namely Pd_2O, PdO are known.

Palladium suboxide : It is prepared by heating palladium at 450°C in oxygen. It can also be obtained by heating palladium monoxide to redness. Pd_2O decomposes with acid into monoxide and metal.

Palladium monoxide or palladous oxide : (PdO). It is obtained by treating solution of palladous salt with NaOH. It may be prepared by fusing palladium with caustic soda. It may be obtained by heating palladium nitrate to redness. It is the most stable oxide of palladium. It is a black powder insoluble in all acids.

Palladium dioxide : (PdO_2) It is obtained by treating a solution of potassium palladate with a slight excess of KOH. The brown precipitate so obtained is washed with water and then dried over conc. H_2SO_4.

PLATINUM

Occurrence

Platinum is mostly found in the native form along with Cu, Au, Fe and other Pt. metals (Os, Ir, Rh, Pd) in sands and gravels. The main deposits of native Pt occur in Russia (Ural mountains), Columbia and

South Africa. In the combined state it occurs in a few minerals like sperrylite, $PtAs_2$; cooperite, $Pt(AsS)_2$ and braggite, (Pt, Pd, Ni) S. Sperrylite occurs in Sudbury district of Canada and Transvaal (South Africa).

Metallurgy

The various steps in the metallurgy of platinum have been as follows:

(i) *Preparation of the concentrate, a crude mixture of the platinum metals.*

(ii) *Separation of platinum from the concentrate.*

(1) Preparation of the Concentrate

(a) From the native metal : After scooping the platiniferous sands and gravels from the river beds they are delivered on to the Wilfley tables where these are washed by the steam of water. Heavier metallic particles get retained behind the riffles while the lighter sand particles get washed away. The concentrate may have 50.85% of platinum together with osmium-iridium. palladium, gold, silver, copper and iron. The concentrate is having two types of grains.

(i) *Crude platinum grains are dull and have platinum, along with some amount of ion copper, gold palladium and iridium. These are soluble in aqua regia.*

(ii) *Osmiridium grains are dull and have mainly osmium and iridium along with small quantities of platinum, palladium, etc. These are soluble in aqua-regia.*

(b) From the Mond process residue : The non-volatile residue remained during refining of nickel by Mond's process is having the platinum along with some copper and nickel. When the roasted residue is made to treat with sulphuric acid when copper and nickel are dissolved out. The undissolved portion is fused with lead which forms an alloy with precious metals. The lead alloy has been subjected to cupellation whereby lead is removed as litharge. Platinum metals are left behind in the concentrate.

(c) From the Orford process muds : In the electrolytic refining of nickel from the 'Orford bottoms' the platinum metals together with gold and silver have been collected as collected as anode mud in the concentrated form.

(2) Separation of Platinum from the Concentrate

When the concentrate obtained by any of the above processes is made to treat with aqua-regia, osmiridium, ruthenium, rhodium and silver (as AgCl) remain in the residue with platinum. Palladium and gold together with some iridium and rhodium go into solution as chloro-complexes (*e.g.*, H_2PtCl_6, H_2PdCl_6, $IrCl_4$, $RhCl_4$ etc).

The solution is then reduced with ferrous sulphate when the chlorocomplex of gold gets reduced to the metal while H_2PdCl_6, $IrCl_4$, and $RhCl_4$ are also mainly reduced to slightly soluble chlorides of the metals in the lower oxidation states ($PdCl_2$ etc.), H_2PtCl_6 is unaffected. The residue left on filtration finds use for the recovery of palladium.

When the filtrate is treated with ammonium chloride, $(NH_4)_2PtCl_6$ alongwith some $(NH_4)IrCl_6$ is precipitated. On igniting the precipitate crude platinum alongwith some iridium remains behind.

$$H_2PtCl_6 + 2NH4Cl \rightarrow (NH_4)2PtCl_6 + 2HCl$$

$$(NH_4)_2PtCl_6 \xrightarrow{\text{Ignition}} 2NH_3 + 2HCl + 2Cl_2 + Pt$$

It is again treated with aqua-regia which dissolves platinum but very little iridium. Platinum is again precipitated as $(NH_4)_2PtCl_6$ which is ignited. The process is repeated until platinum becomes free from iridium.

Properties

Its colour appears as white as tin. With the exception of gold and silver it is most malleable of the metals. It joins well with glass as its coefficient of expansion is the same as that of glass. At red heat temperature, it easily allows hydrogen to pass through it. The pure metal is not attacked by hot concentrated sulphuric acid. It however, dissolves in hot aqua-regia. The alloy of lead and platinum dissolves in nitric acid producing platinum nitrate.

Various TTypes of Platinum

Platinum may be formed in different forms in different forms to be used in different industries.

1. **Spongy platinum :** If the double salt ammonium platinochloride $(NH_4)_2PtCl_6)$ is very gently heated, a type of grey porous form having a large surface is produced which is called spongy

platinum. It is having great surface activity and is used for adsorption of gases.

2. **Platinum black :** (a) If an alloy of platinum with metals such as copper or zinc is reacted with nitric acid, the nascent reduces the platinum salt and falls back as a black powder. (b) It is best produced by reducing platinic chloride with formaldehyde, or reducing neutral chloroplatinate with sodium formate. The platinum black is known to adsorb much more hydrogen than the spongy platinum (160 volumes), the maximum adsorption is at 0°C; and thus, it is commercially used in reduction processes. It also serves as a good catalyst. Alcohol in oxidized by it because of its occluded oxygen. In its presence, a mixture of hydrogen and oxygen explodes.

3. **Platinized asbestos :** If asbestos, soaked with platinic chloride, a little hydrochloric and some ammonium chloride, is made to heat in a crucible, or if reduced with sodium formate, platinum precipitates out in the fibers of the asbestos. This is now called platinized asbestos and extensively used as catalyst.

4. **Colloidal platinum :** A colloidal solution of platinum can be obtained by Bredig's arc method. Spongy platinum black, and platinized asbestos adsorb large amounts of oxygen gas (100-110 times their volume) and then act as oxidizing agents.

Properties

It is heavy soft, ductile and malleable metal having a strong silvery white luster. It can be sealed into glass as the two have approximately same coefficient of thermal expansion. It is noble metal and is only slightly affected even on prolonged heating in air. The fused metal dissolves oxygen that is given out suddenly on cooling. The phenomenon is called spitting. It is quite resistant to the action of acids. Boiling concentrated sulphuric acid attacks it to a slight extent. Aqua-regia (3 parts conc. HCl and 1 part conc. HNO_3), however, dissolves it readily forming chloroplatinic complex.

$$3HCl + HNO_3 \rightarrow 2H_2O + NOCl + 2Cl$$

$$Pt + 4Cl \rightarrow PtCl_4$$

$$PtCl_4 + 2Cl^- \rightarrow [PtCl_6]^{2-}$$

A solution of a chlorate in the presence of concentrated hydrochloric acid also dissolces it readily.

With fluorine and chlorine it gives the tetrahalides at 773K and 1773 K respectively. Many metals (*e.g.*, Pb, Sn, etc.) and C, P and As are known to form alloy with platinum, hence platinum vessels should not be heated in a luminous flame and in contact with these elements or their compounds in presence of carbon. Many substances in the fused state (e.g. alkali hydroxides, nitrates and bisulphates) and CS_2 vapours also attack platinum. Hence such substances are not to be heated in platinum crucibles and platinum vessels should not be heated in a coal gas flame containing small quantities of carbon disulphide vapours.

Uses

1. *As it is not attacked by air, oxygen or water, it is used in jewellery and because of its bright lustre, it is used for ornamental purposes. It is also used in dentistry.*
2. *Since it is not attacked by strong acids even at high temperatures, it is used in chemical apparatus like crucibles, dishes, etc. It is also used as electrodes in various electrolytic processes.*
3. *As it can be drawn into extremely thin wires and foils, it is used making a number of instruments needed for various scientific purposes.*
4. *It is employed as an important catalyst in the form of platinised asbestos, that is prepared by soaking asbestos in a solution of platinic chloride having a little of hydrochloric acid and ammonium chloride. Some important reaction catalysed by platinum have been as follow*

(i) *Reaction of sulphur dioxide with oxygen to form sulphur trioxide in the contact process of the manufacture of sulphuric acid.*

(ii) *Oxidation of ammonia to nitric oxide in the Ostwald's process for the manufacture of nitric acid.*

(iii) *Oxidation of methyl alcohol to formaldehyde.*

(iv) *Decomposition of* N_2H_4 *to* N_2 *and* NH_3.

SELENIUM

History

In 1817 Berzelius discovered this element from the red deposit obtained from the chamber of sulphuric aid plant. He named the element as selenium (in Greek, selene means moon).

Occurrence

Native state : It is occasionally met within the free state in native sulphur, particularly Japanese sulphur. A large number of metallic sulphides are associated with small quantities of selenium.

Combined state : It occurs in number of minerals mostly selenides and few selenites.

Selenolite	SeO_2
Zorgite	Cu_2Se, PbSe - 31% Selenium
Crookesite	$(Cu, Al, Ag)_2Se$
Berzelianite	Cu_2Se
Gua Sajuatite	Bi_2Se_3
Chalcomenite	$CuSeO_3.2H_2O$

EXTRACTION

A number of methods have been employed for the extraction of element selenium.

1. From Anode Mud

Anode mud is obtained during the electrolytic refining of copper. The mud contains the particles of gold and silver with the impurity of selenium. Selenium may be obtained from this mud in the following two ways :

(a) Oxidation Method

When the mud is roasted under oxidising conditions, copper (if present) and selenium are converted into their corresponding oxides. Selenium dioxide is volatilised and is deposited in the flues. Selenium dioxide so obtained is dissolved in water and then reduced with sulphur dioxide to precipitate selenium.

$$SeO_2 + H_2O \rightarrow H_2SeO_3$$

$$H_2SeO_3 + 2SO_2 + H_2O \rightarrow Se + 2H_2SO_4$$

After filtering off, the gray coloured ppt. of selenium is washed and dried.

The main drawback of this method is that a small amount of selenium and tellurium remains unaffected by roasting and is thus not isolated. This residual selenium can be extracted by the method (a).

(b) Alkali Fusion Method

When the anodic mud or the residue which escapes roasting is fused with sodium hydroxide and sodium nitrate, selenium and tellurium form selenite and tellurite respectively.

$$Se + 2NaOH + 2NaNO_3 \rightarrow Na_2SeO_3 + 2NaNO_2 + H_2O$$

$$Te + 2NaOH + 2NaNO_3 \rightarrow Na_2TeO_3 + 2NaNO_2 + H_2O$$

When the fused mass is extracted with water and then acidified with HCl, tellurium is precipitated out as oxide while selenium remains in solution as selenium acid.

$$Na_2SeO_3 + 2HCl \rightarrow H_2SeO_3 + 2NaCl$$

$$Na_2TeO_3 + 2HCl \rightarrow TeO_2 + H_2O + 2NaCl$$

The contents are filtered and the treated with sulphur dioxide when selenium is precipitated as a gray powder.

$$\text{Anode mud} \xrightarrow{\text{Roasted}} \text{Roasted mud} \xrightarrow[\text{NaOH + NaNO3}]{\text{Heated with}} \text{Melt}$$

Extracted

with water

$$\text{Selenium} \xleftarrow[\text{Passed}]{\text{So2}} H_2SeO_3 \xleftarrow[\text{with Hcl}]{\text{Acidified}} \text{Water extract}$$

(Na_2SeO_3) and (Na_2TeO_3)

2. From Flue Dust

Selenium is generally associated with iron pyrites. It gets converted into SeO_2 during roasting of iron pyrites. Selenium dioxide passes along with sulphur dioxide into the chamber where it is collected as a red coloured dust. This is commonly known as flue dust and is treated in several ways for getting pure selenium.

(a) Potassium Permaganate Method

When the flue dust is agitated with dry powdered potassium permanganate in the presence of sulphur dioxide at about 50-60° C, selenium is oxidised to selenium dioxide. To this solution sodium chloride is added to remove any silver as silver chloride. The contents are diluted and filtered. The filtrate is acidified and SeO_2 is reduced by passing SO_2 gas to metallic selenium.

$$4KMnO_4 + 6H_2SO_4 + 5Se \rightarrow 5SeO_2 + 4KMnSO_4 + 2K_2SO_4 + 6H_2O$$

$$SeO_2 + 2SO_2 + 2H_2O \rightarrow Se + 2H_2SO_4$$

(b) Chlorine Method

When flue dust is heated in a current of chlorine, selenium and sulphur form their corresponding chlorides. The vapours of chlorides are absorbed in water when $SeCl_4$ is converted into selenious acid. The solution so formed is acidified with HCl to form SeO_2 which is then reduced with SO_2 to form metallic selenium.

$$SeCl_4 + 3H_2O \rightarrow H_2SeO_3 + 4HCl.$$

(c) Sulphuric acid Process

The flue dust is ground to a fine powder and digested with fuming sulphuric acid. The oxide formed (SeO_2) is sublimed or recrystallised as selenious acid (H_2SeO_3). This is reduced by SO_2.

$$Se + 2SO_3 \rightarrow SeO_2 + 2SO_2$$

$$SeO_2 + H_2O \rightarrow H_2SeO_3$$

$$H_2SeO_3 + 2SO_2 + H_2O \rightarrow Se + 2H_2SO_4$$

(d) Potassium Cyanide Process

Finely powdered flue dust is digested with KCN when selenium dissolves to form potassium selenocyanate (KCNSe) and on adding HCl selenium is precipitated as a red powder. It is purified by evaporation with HNO_3 to dryness, yielding SeO_2. This is recrystallised as selenious acid. This is reduced to selenium by passing SO_2 through the solution.

$$Se + KCN \rightarrow KCNSe$$

$$KCNSe + HCl \rightarrow KCl + HCN + Se$$

$$Se + 2HNO_3 \rightarrow SeO_2 + NO + NO_2 + H_2O$$

$$SeO_2 + H_2O \rightarrow H_2SeO_3$$

$$H_2SeO_3 + 2SO_2 + H_2O \rightarrow 2H_2SO_4 + Se$$

Purification : Selenium formed from above may contain copper, lead, etc, as impurities. The crude element is reacted with nitric acid and evaporated to dryness, with the result solid selenium dioxide is formed. During the process for crystallization from hot water, impurities remain in the mother liquor and selenious acid H_2SeO_3 is obtained. A solution of this acid in water is then reduced by passing SO_2 to obtain the pure Se as a red powder.

Allotropic Forms of Selenium

Selenium occurs in the following four forms :

(a) Amorphous Selenium

It is obtained by the reduction of selenium compounds with sulphuric acid. It is a bright red powder.

(b) Vitreous Selenium

It is obtained by heating amorphous selenium above 217° C and then cooling it rapidly. It forms black glassy mass which can be electrified by friction. Its specific gravity is 4.5. It consists of Se molecules of various lengths.

Properties

Physical : Selenium is greyish solid with metallic cluster. It occurs in allotropic forms like sulphur as amorphous selenium, vitreous selenium, red monoclinic selenium (crystalline) and grey selenium (metallic). The sepecific gravity varies according to selenium forms ranging between 4.3 to 4.8. Selenium melts at 217°C and boil at 685°C giving off dark red powder.

Chemical : Pure selenium is stable in air and does not form oxide at ordinary temperature. It resembles sulphur and tellurium in many chemical properties.

(a) **Reaction with hydrogen :** At high temperature selenium combines with hydrogen directly forming hydrogen selenide.

$$Se + H_2 \xrightarrow{250-500° C} H_2Se$$

(b) **Reaction with oxygen :** When heated in air or oxygen, selenium burns with a bright blue flame forming SeO_2 giving unpleasant smell.

$$Se + O_2 \longrightarrow SeO_2$$

(c) **Reaction with water :** It reacts with water vapours forming selenic acid (in presence of ozone as oxidising agent).

$$Se + H_2O + O_3 \rightarrow H_2SeO_4$$

H_2O_2 has similar effect and forms H_2SeO_4.

(d) **Reaction with halogens :** Selenium directly combines with halogens especially fluorine with which it forms a white fluoride (gas and liquid)

$$Se + 2F_2 \rightarrow SeF_4$$
$$Se + 3F_2 \rightarrow SeF_6$$

Chlorine reacts with Se forming Se_2Cl_2 and $SeCl_4$.

$$2Se + Cl_2 \rightarrow Se_2Cl_2 \text{ (Selenium monochloride)}$$

Se_2Cl_2 changes into $SeCl_4$ on heating.

$$2Se2Cl_2 \rightarrow 3Se + SeCl_4$$

Similarly selenium combines with bromine and iodine to yield similar compounds.

(e) **Reaction with acid :** Selenium reacts with H_2SO_3 and also with HNO_3.

$$Se + H_2SO_3 \rightarrow SeSO_3 + H_2O$$

On dilution selenium is precipitated.

$$SeSO_3 + H_2O \rightarrow H_2SO_4 + Se$$
$$Se + 2HNO_3 \rightarrow SeO_2 + H_2O + NO + NO_2$$
$$SeO_2 + H_2O \rightarrow H_2SeO_3$$

(f) **Reaction** with metals : Selenium combines with many metals to form selenides and alkali metal selenides and polyselenides (Ne_2Se_3, Na_2Se_4, Na_2Se_5).

In fused state it combines with antimony, lead, bismuth, copper, silver and gold.

Compounds of Selenium

Selenium compounds show close resemblance to sulphur compounds. The compounds of selenium are divalent, tetravalent and hexavalent

Hydrogen selenide : It is prepared by the action of nascent hydrogen on selenious acid.

$$H_2SeO_3 + 6H \rightarrow H_2Se + 3H_2O$$

H_2Se can also be obtained by reacting iron or sodium selenide with HCl.

$$FeSe + 2HCl \rightarrow FeCl_2 + H_2Se$$

$$Na_2Se + 2HCl \rightarrow 2NaCl + H_2Se$$

Properties

(a) it is a colourless and inflammable gas, with b.p. –41.7°C and m.p. –64°C. It has unpleasant small and is poisonous. It is relatively less stable than H_2S and oxidises on exposure to air.

(b) Its aqueous solution absorbs oxygen from air with the separation of selenium.

$$2H_2Se + O_2 \rightarrow 2Se + 2H_2O$$

(c) It burns in air or oxygen forming water and selenium.

$$2H_2Se + O_2 \rightarrow 2Se + 2H_2O$$

(d) It burns in air or oxygen forming water and selenium.

$$H_2Se + S \rightarrow H_2S + Se$$

Selenium dioxide : selenium dioxide is prepared by heating selenium in oxygen.

Properties

(a) SeO_2 is a white crystalline, volatile oxide.

(b) SeO_2 has a specific gravity of 8.95 and sublimes on heating at 310°C.

(c) It is easily soluble in water forming selenious acid H_2SeO_3 of which it is the anhydride.

In the gaseous form SeO_2 has the same structure as that of SO_2. However, its solid form consists of non-planar chains.

Selenium dioxide is used to oxidize aldehydes and ketones.

$$R - CH_2 - CO - R + SeO_2 \rightarrow R - CO - CO - R + H_2O + Se$$

Selenium trioxide : it is prepared by treating a solution of selenium or $SeOCl_2$ with ozonised oxygen.

$$Se + O_2 \rightarrow SeO_2$$

$$SeOCl_2 + 2O_3 \rightarrow SeO_3\ 2O_2 + Cl_2$$

It may also be formed by passing a glow discharge through a mixture of oxygen and selenium vapours at 40°. It is a white, deliquescent solid soluble in water yielding acidic solution.

$$SeO_3 + H_2O \rightarrow H_2SeO_4$$

It decomposes at 120° C into selenium dioxide and oxygen.

$$2SeO_3 \rightarrow 2SeO_2 + O_2$$

Selenium hexafluoride : This is prepared by the action of selenium of fluorine kept at –78°.

$$Se + 3F_2 \rightarrow SeF_6$$

Properties

It is white crystalline substance at –78° C. SeFe6 has specific gravity of 2.51. The m.p. and subl. pt. are –35 and –6° C respectively.

Selenium tetrafluoride : It is obtained by the action of AgF on Selenium tetrachloride.

$$4AgF + SeCl_4 \rightarrow SeF_4 + 4AgCl$$

It can also be obtained by the action of fluorine on selenium at ordinary temperature.

$$Se + 3F_2 \xrightarrow{78°C} SeF_4$$

Properties

It is a colourless liquid with a specific gravity of 2.77. Its m.p. and b.p. are –13 and 93° C respectively. It is easily hydrolysed by water.

$$SeF_4 + 2H_2O \rightarrow SeO_2 + 4HF$$

Selenium tetrachloride, $SeCl_4$: It is prepared by the action of chloride of chlorine on selenium monochloride or by treating selenium oxide with phosphorus pentachloride.

$$Se_2Cl_2 + 3Cl \rightarrow 2SeCl_4$$

$$6SeO_2 + 6PCl_5 \rightarrow 6SeCl_4 + P_4O_{10} + 2POCl_3$$

It is yellow solid which is readily hydrolysed by water.

Selenium monochloride, Se_2Cl_2 : It may be prepared by passing hydrogen chloride into a solution of selenium in fuming nitric acid or by passing a current of chlorine over heated selenium. It is a brown coloured, oily liquid which decomposes on heating into selenium and selenium and selenium tetrachloride. It is readily hydrolysed by water to yield selenious acid.

Selenious acid : It is an oxy-acid of selenium and is prepared by the evaporation of SeO_2 in water or nitric acid.

$$SeO_2 + H_2O \rightarrow H_2SeO_2$$

$$Se + 2HNO_3 \rightarrow H_2SeO_2 + NO + No_2$$

Properties

1. *It is a white crystalline substance. It loses water on heating. Selenious acid and selenites are and selenites are easily reduced to red selenium by organic matter.*

$$H_2SeO_3 \rightarrow H_2O + SeO_2$$

$$\underset{Org}{SeO_2 + 2H} \rightarrow \underset{Org}{Se + 2OH}$$

2. *He_2SeO_3 forms acidic and normal selenites $KHSeO_3$ and K_2SeO_3.* $H_2SeO_3 + KOH \rightarrow KHSeO_3 + H_2$

3. *It forms heteropoly acids with vanadic, molybdic and uranic acid.*

Selenic acid : It is another oxy-acid of selenium. It is prepared by oxidation of Selenious acid or selenium by oxidising agents such as $KMnO_4$, chlorine or bromine water.

$$H_2SeO_3 + Cl_2 + H_2O \rightarrow H_2SeO_4 + 2HCl$$

$$Se + 4H_2O + 3Cl_2 \rightarrow H_2SeO_4 + 6HCl$$

It can also be prepared by the action of bromine on silver selenite.

$$Ag_2SeO_3 + H_2O + Br_2 \rightarrow H_2SeO_4 + 2AgBr$$

Properties

It is a white solid with specific gravity of 2.95. Its m.p. and b.p. are 58°C and 265°C respectively. If forms crystalline hydrates containing 1 and 4 molecules of water. The acid is extremely hygroscopic and evolves heat with water.

Dilute H_2SeO_4 dissolves zinc, barium, calcium and some iron.

$$Zn + H_2SeO_4 \rightarrow ZnSeO_4 + H_2$$

Selenic acid is reduced to selenious acid by boiling with dilute HCl or reduced by H_2S and SO_2 in sulphur and Se.

$$H_2SeO_4 + 2HCl \rightarrow H_2SeO_3 + H_2O + Cl_2$$

Uses of Selenium

1. **In glass industry :** Selenium decolourises the green tinge of the glass due to ferrous ion.

The green colour is complemented with the pink colour produced by selenium. It is also used to produce a ruby glass which is used for signal lights.

2. ***Rubber industry*** *: It improves the resistance of rubber to friction . Such a vulcanised rubber affords a fire proof material for cables. It makes for cables. It makes rubber less liable to flame.*
3. ***Selenium cells*** *: Since selenium is very sensitive to light, it is used the construction of selenium calls which have important applications such as electrical transmission of photographs, burglar alarms, electronics, etc.*
4. ***Optophone*** *: It is used in optophone (used by a blind man to read ordinary print). It consists of a selenium cell a telephone and illuminating device.*
5. ***Pigments*** *: Certain selenium compounds such as barium and lead selenides have been employed as pigments because of their fine grains, white colour and intense capacity .*
6. *Colloidal selenium is used as germicide, insecticide and fungicide.*
7. *Se is used in the production of stainless steel in which it is added from 0.03 to 2.0 per cent. Selenium in the form of Se-S mixture is used in protecting workers from mercury poisoning.*
8. *It is widely used in the manufacture of special dyes.*
9. *Diethyl selenide has been suggested as an antiknock agent for petrol.*
10. *In petroleum industry iron selenides have been used as a catalyst in the cracking of petroleum.*

CERIUM

Introduction : Cerium is one of the rare-earth elements and is included in that group of elements. Though, it is in third group, its quadrivalency is well-established and in many ways it resembles thorium.

History

In 1803, the mineral cerite was analysed by ***Klaproth*** and ***Berzelius*** and it was found that mineral contained the new oxide ceria similar to yttria. Actually the oxide was found to contain later on a mixture of may rare-earth oxides. The metal cerium was isolated in 1826.

Occurrence

Cerium is most abundant of rare-earth metals. It occurs associated with other elements like iron, thorium etc.

1. **Euxenite :** It consists of niobates and tantalates of cerium earth.
2. **Bastnaesite :** It consists of fluocarbonate of cerium earth together with traces of thorium.
3. **Cerite :** It is hydrated silicate of cerium, $H_3(Ca,Fe)Ce_3Si_3O_{13}$, Ce: 51— 72%, Found in Sweden and Caucasus.
4. **Monazite :** It is a phosphate of cerium earth containing thorium and silica, Ce : 49—74%. Found in India, Brazil, Carolinas, Idaho and Scandinavia.
5. **Aeschnite :** It is a titano-niobate of cerium with thorium, Ce : 19.4—24.1%. Found in Norway, Germany, Brazil.
6. **Allanite :** It is hydrated silicate of cerium earth containing calcium, iron, beryllium, aluminium, Ce : 3.6 —51%. Found in Greenland and Scandinavia.

Extraction

For cerium monazite is the best source. Separation of cerium from other rare-earth elements, is based upon the fact that unlike other rare-earth elements cerium forms two distinct series of compounds derived from strongly basic $Ce(OH)_3$ and weakly, basic $Ce(OH)_4$. Ce(III) can be oxidised to Ce(IV) by means of $KBrO_3$., $KMnO_4$ etc which on boiling gives insoluble basic salt. The precipitation is fairly complete and adsorbs other rate-earths.

Monazite sand is an important source for the extraction of cerium. Monazite sand is treated with concentrated sulphuric acid. The reaction mixture is then lixivated with water and filtered. The rare-earths are converted into water soluble anhydrous rare-earth sulphate except thorium which is separated as insoluble residue. Part of thorium gone into the filtrate may be precipitated by the addition of sodium pyrophosphate and filtered out.

The filtrate is treated with oxalic acid to get the precipitates of rare-earths as oxalates. The precipitate is isolated by filtration and then boiled with alkali hydroxide the granular rare-earth hydroxides are formed which are washed and filtered.

Cerium can be very readily separated from the other rare-earth by oxidation of Ce(III) to Ce(IV). Since ceric hydroxide is the weakest base of all and gets therefore, precipitated easily with dilute alkalies of weak bases.

The Ce(III) is oxidised in Ce(IV) with potassium permangnate, potassium persulphate, potassium bromate, chlorine, hydrogen peroxide or a mixture of nitric acid and ammonium nitrate as below :

1. Permanganate Method : In this method a neutral solution of nitrates is heated upto boiling and $KMnO_4$ is added till a pink colour appears. Cerium is then precipitated together with manganese, by adding the reagent consisting of $KMnO_4$ and Na_2CO_3 in the molar ratio.

$3Ce(NO_3)_3 + KMnO_4 + 4Na_2CO_3 + 8H_2O \rightarrow 3Ce(OH) \downarrow + MnO_2.2H_2O + 8NaNO_3 + KNO_3 + 4CO_2 \downarrow$

2. Bromate Method : The concentrated neutral solution of nitrates of rare earths is obtained by treating the oxide mixture of rare-earths with HNO_3. The solution of the nitrates is heated to boiling and potassium bromate is added. As soon as bromine fumes appear, lumps of marble are added and boiling is continued for 2-3 hours. When precipitate of basic cerium nitrate $[Ce(OH)(NO_3)_3]$ is obtained, boiling should be stopped. The marble is then removed and precipitate is allowed to settle. The precipitate adsorbs some other members of the rare-earth groups. The separated basic salt precipitate is again dissolved in HNO_3 and separation by bromate method is repeated once or twice.

3. Chlorine Method : In this method, Cl_2 is passed through the suspension of the hydroxides in an excess of caustic soda solution when $Ce(OH)_4$ remains undissolved while the other hydroxides dissolve as chlorides and hypo-chlorites. Several repetitions produce a satisfactory separation.

4. Ceric Ammonium Nitrate Method : The basis of the method is that the double salt $Ce(NO_3)_4 2NH_4NO_3$ is much less soluble in HNO_3 than the corresponding salts of other rare earth-metals.

In this method the mixed oxides are dissolved in nitric acid and calculated quantity of ammonium nitrate is added to the hot solution. On cooling, the double salt-ceric ammonium nitrate separates as orange yellow crystals. It is purified by recrystallisation.

5. Electrolytic Method : It is the most efficient and rapid method of separating cerium from rare-earths. The solution of nitrates or sulphates is mixed with sodium phosphate and electrolysed when ceric phosphate is precipitated at the anode.

6. Solvent Extraction : A more recent method of separation consists in *solvent extraction* of ceric solutions in the tetrapositive in the tetrapositive valence state using tributyl phosphate (TBP) in mineral oil as the organic solvent.

The organic phase is stripped of its cerium content by reduction of Ce(IV) to Ce(III) by means of H_2O_2. The resulting Ce(III) compounds, being insoluble in the organic phase, passes into the aqueous phase.

Cerium From Reactor Fuel

An additional source of cerium now-a-days is barium-140, recovered from used up reactor fuels. It is done by treating the latter with NaOH solution, which dissolves aluminium, leaving uranium and most of the fission products in the residue.

The insoluble portion is dissolved in HNO_3 and the solution concentrated to small bulk. Fuming HNO_3 is now added, when barium gets precipitated as barium nitrate which contains some strontium nitrate as contaminant.

The impure salt is dissolved and the solution treated with K_2CrO_4 at controlled pH, when $BaCrO_4$ gets precipitated, while strontium is left behind in solution. Repetition of the procedure gives a fairly pure product.

Barium- 140 obtained from the nitrate, is β-active and undergoes the following decay chain to produce cerium-140.

$$^{140}_{56}\text{Ba} \xrightarrow[12.8 \text{ days}]{-\beta} {}^{140}_{57}\text{La} \xrightarrow[40.2 \text{ hrs.}]{-\beta} {}^{140}_{58}\text{Ce (stable)}$$

Metallic Cerium

(a) Metallic cerium was first prepared by reduction of ceric chloride with sodium.

$$CeCl_4 + 4Na \rightarrow Ce + 4NaCl$$

But the yield is poor and is contaminated with NaCl.

Similarly, reduction of fluoride with calcium or aluminium yields an impure metal.

(b) The metal is commercially prepared by the electrolysis of a fused mixture of chlorides of cerium, calcium and sodium. The mixture is melted in an iron pot, which is made the cathode during the electrolysis.

The anode is an iron or carbon rod. Cerium in the molten condition, collects at the bottom of the pot, it is trapped and cast into plates.

Properties

Physical

(a) Cerium is steel-grey, lustrous metal.

(b) It is malleable and ductile and is soft enough to cut with a knife.

(c) It is a poor conductor of electricity.

(d) It exists in four allotropic modifications.

(e) It produces sparks when struck with a piece of steel.

(f) It is paramagnetic.

(g) It melts at 795° C.

Chemical

Chemically, cerium is quite active.

1. Action of Air : Cerium is stable in dry air but it tarnishes in most air. It has low ignition temperature and catches fire at 160°C. On burning it produces ceric oxide (CeO_2) with much heat of formation.

2. Action of Water : Cold has little action on cerium but cerium evolves H_2 from boiling water.

$$2Ce + 6HCl \rightarrow 2Ce(OH)_3 + 3H_2$$

3. Action of acid : Dilute and concentrated HCl and HNO_3 attack the metal at room temperature.

$$2Ce + 6HCl \rightarrow 2CeCl_3 + 3H_2$$

$$3Ce + 3HNO_3 \rightarrow 3CeO_2 + 4NO + 2H_2O$$

Only dilute H_2SO_4 attacks the metal while concentrated H_2SO_4 has practically no action.

$$2Ce + 3H_2SO_4 \rightarrow Ce_2(SO_4) + 3H_2$$

4. Combination with Non-metals : Cerium directly combines with H_2 at 345°C and forms bluish-black hydride which is readily decomposed at high temperatures. The hydride is of interstitial type.

It burns with Cl_2 at 220°C forming its chloride, $CeCl_3$. It also combines with I_2, S, As, Se, Te, P, C, N_2, etc. at different temperatures to form the binary compounds.

5. Reducing Action : Cerium reduces CO, CO_2 and CCl_4, with liberation of carbon. For this reason, CO_2 and CCl_4 are ineffective in putting out cerium fires.

Uses

1. Pure metallic cerium has no commercial uses but in the form of its alloys it has interesting applications. The important alloy of cerium is *misch metal.* It is used in gas and cigarette lighters.
2. Addition of a small proportion of cerium or misch metal to stainless steel improves their workability remarkably.
3. Addition of misch metal to steel has been found to increase the strength as much as 25 percent.
4. Magnesium alloys containing misch metal (3%) and zirconium (1%) having sufficiently high strength at elevated temperatures have been found particularly useful in making jet engine parts.
5. Cerium carbide and nitride are used in arc electrodes and in filaments for incandescent electric bulbs.
6. Cerium salts are used as a catalyst in the oxidation of hydrogen chloride to chlorine. Cerium hydride and nitride are used in the combination of N_2 and H_2 to form NH_3.
7. Ceric salts being oxidising agents are used in the manufacture of aldehydes, quinones, etc. from aromatic hydrocarbons.
8. Ceric salts are used as a mordant in dyeing cotton.
9. It is also used in glass for spectacles.

Detection

(i) *Small quantities of cerium can be detected by adding the neutral solution to a warm concentrated potassium carbonate solution and then adding a few drops of dilute hydrogen peroxide. A yellow colour is produced.*

(ii) *If the solution of a cerous salt when treated with ammonium hydroxide, an intense red-brown precipitate is formed.*

(iii) *A solution of a ceric salt when treated with ammonium hydroxide or sodium hydroxide yields yellowish precipitate.*

(iv) *A very sensitive test for cerium (III) is provided by an ammoniacal silver nitrate. The reagent is added to a neutral solution, when a mixed precipitate, consisting of Ce(IV) hydroxide (Yellow) and finely divided metallic silver (black) is formed.*

$$Ce^{3+} + [Ag(NH_3)_2]^+ + 4OH^- \rightarrow Ce(OH)_4 + Ag + 2NH_3$$

The test should be performed in the absence of Fe, CO *and* Ni.

(v) ***Dry test :*** *When a cerium is fused in to a borax bead in an oxidising flame, the bead is dark brown when hot and pale yellow to colourless when cold. In the reducing flame, the bead becomes colourless.*

Spot Tests

(a) Test For Ce^{3+} Compounds

(i) ***Phosphomolybdic acid test :*** *A drop of the test solution and of phosphomolybdic acid solution are mixed on a spot plate, and a drop of 40% sodium hydroxide is added. A intense blue colour or precipitate appears.*

(ii) ***Ammonaical silver nitrate test :*** *As described above in test number in detection (iv).*

(b) Test For Ce^{4+} Compounds

(i) ***Benzidine test :*** *A drop of the test solution is placed on filter paper, followed by a drop of dilute sodium hydroxide, and by a drop of benzidine solution. A deep blue colour develops (Blazejak-Ritges, 1970).*

(ii) **4′. *Amino acetophenone test :*** *This reagent reacts with* Ce^{4+} *compounds in 1.5 M HCl or* H_2SO_4 *solution giving a purple colour which changes to a red-brown precipitate (Somidevamma 1947).*

Estimation

(i) ***Gravimetric estimation :*** *In a ceric salt, cerium can be estimated gravimetrically by precipitation as ceric iodate,* $Ce(IO_3)_4$*, by using a solution of* KIO_3*. The precipitate is boiled with water and oxalic acid until all iodine has been expelled.*

It is then filtered, washed, dried, ignited and weighed as cerium dioxide, CeO_2*. In the case of cerous salts, cerous is converted into ceric by oxidation with potassium bromate and concentrated nitric acid, before precipitation as the iodate.*

(ii) ***Volumetric estimation :*** *Cerium in a ceric salt can be readily estimated by adding to a measured volume of the ceric solution acidified with sulphuric acid, N/20 ferrous ammonium sulphate solution in sufficient excess (as shown by complete discharge of yellow colour of ceric ion and back titrating the excess of ferrous ion present by means of standard* $KMnO_4$*).*

In the case of a cerous salt, the solution is first submitted to oxidation to the ceric state by means of sodium bismuthate in the presence of sulphuric acid. A measured volume of the oxidised solution is then subjected the above procedure.

Compounds

Cerium is known to form two classes of compounds. Cerous compounds, in which the oxidation number of cerium is +3, by the loss of there electron *i.e.,* two from 4f-orbitals; hence, the electronic configuration of Ce(III) ion would be $[Kr]4f^1$. Ceric compounds, in which the oxidation number of cerium is +4, by the loss of four electrons, *i.e.,* two from 6s and two from 4f-orbitals; hence the electronic configuration of Ce(IV) ion is [Kr].

Cerous Oxalate $Ce_2(C_2O_4)_3.10H_2O$

Preparation

It is obtained when oxalic acid is added to solution of cerous salt.

$$2Ce(NO_3)_3 + 3H_2C_2O_4 \rightarrow Ce_2(C_2O_4)_3 + 6HNO_3$$

Properties

It is a white crystalline powder, insoluble in water. It dissolves only to a slight extent in dilute strong acids, a behaviour that is typical of all the rare-earth oxalates.

Uses

It finds use medicinally as a sedative in sickness, particularly that of pregnancy, also in chronic-diarrhoea, hysteria, epilepsy and migraine.

Ceric Sulphate $Ce(SO4)_2$

Properties

It is obtained by heating a mixture of ceric oxide and concentrated sulphuric acid on a sand bath. The product is washed with ice-cold water and dried over caustic alkali.

$$CeO_2 + 2H_2SO_4 \rightarrow Ce(SO)_2 + 2H_2O$$

Properties

It is a deep orange yellow crystalline powder It decomposes on heating at above 195°C. It dissolves in water yielding a brown-yellow solution from which salt separates below 3°C as the do decahydrate, between 3° and 42.5° as the octahydrate and above 42.5°C as the tetrahydrate.

On diluting an aqueous, solution of ceric sulphate hydrolysis takes place and ultimately, sparingly soluble basic salts get precipitated.

It is a good oxidising agent for the oxidation of potassium nitrite, formic acid, acetic acid, tartaric acid, oxalic acid, methyl alcohol, hydrazine, hydroxylamine etc.

$$2Ce(SO_4)_2 + H_2C_2O_4 \rightarrow Ce_2(SO_4)_3 + H_2SO_4 + 2CO_2 \uparrow$$

$$2Ce(SO_4)_2 + 2N_2H_4 \rightarrow Ce_2(SO_4)_3 + (NH_4)_2SO_4 + N_2 \uparrow$$

$$2Ce(SO_4)_2 + Na_2S_2O_3 \rightarrow Ce_2(SO_4)_3 + Na_2SO_4 + Na_2S_4O_6$$

It forms double salts as follows :

$$2K_2SO_4.Ce(SO_4)_2.2H_2O ; 3(NH_4)_2SO_4.\ xCe(SO_4)_2.\ yH_2O.$$

Uses

It finds use it volumetric analysis as an oxidising agent *e.g.,* for the titration of Fe^{2+}, $C_2O^{2-}{}_4$, $NO^-{}_2$, Cu^+, Mo^{5+}. U^{4+} and Te^{4+} ions.

Ceric Oxide, CeO_2 : It is prepared by heating cerous or ceric hydroxide, nitrate, sulphate and oxalate.

$$Ce(C_2O_4)_2 \rightarrow CeO_2 + 4CO \uparrow + O2 \uparrow$$

It is a white powder having yellowish tint. It turns to lemon-yellow on heating. It is stable oxide but vapourizes at 900°C in vacuum and melts at the temperature of the electric furnace. It is insoluble in acids except concentrated sulphuric acid.

$$CeO_2 + 2H_2SO_4 \rightarrow Ce(SO)_2 + 2H_2O$$

It is a very weak basic oxide. It reacts with uranium dioxide yielding deep blue cubic crystals of $UO_2.CeO_2$.

Uses

It is used in highly efficient glass polishing compositions.

Ammonium Hexanitratocerate (IV), $(NH_4)[Ce(NO_3)_6]$

Preparation

This is commonly known as ceric ammonium nitrate. It is obtained by evaporating a mixed solution of ceric hydroxy nitrate, $Ce(OH)NO_3)_3$. $4\frac{1}{2}H_2O$ and ammonium nitrate in concentrated nitric acid.

$$Ce(OH)(NO_3)_3 + 2NH_4NO_3 \xrightarrow{HNO_3} (NH_4)_2[Ce(No_3)_6]$$

Properties

The Compound is a red crystalline solid. It is hydrolysed in water giving hydrosol and hydrogen. In acidic medium it is a strong oxidising agent. In the reduction of ceric salts in acid solutions with reducing agents, the single valency change occurs

$$Ce^{4+} + e \rightleftarrows Ce^{3+}$$

The equivalent weight of a ceric salt is therefore the molecular weight. In many volumetric redox titrations ammonium hexanitratocerate in the presence of H_2SO_4 can be employed for which ceric sulphate is used.

Ammonium hexanitrato oxidises H_3AsO_3 to H_3AsO_4, Fe^{2+} to Fe^{3+}, $[Fe(CN)_6]^{4+}$ to $[Fe(CN)_6]^{3-}$.

$$2Ce^{4+} + H_3AsO_3 + H_2O \rightarrow 2Ce^{3+} + H_3AsO_4 + 2H^+$$

$$Ce^{4+} + Fe^{2+} \rightarrow Ce^{3+} + Fe3^+$$

$$Ce^{4+} + [Fe(CN)_6]^{4-} \rightarrow Ce^{3+} + [Fe(CN)_6]^{3-}$$

It also oxidises oxalic acid to CO_2.

$$H_2C_2O_4 + 2Ce^{4+} \rightarrow 2CO_2 + 2Ce^{3+} + 2H^+$$

Potassium nitrite is oxidised to KNO_3.

$$KNO_2 + 2Ce^{4+} + H_2O \rightarrow 2Ce^{3+} + KNO_3 + 2H^{+}$$

INDIUM

Discovery

In 1863 indium was discovered by? *Reich* and *Richter*, while investigating sphalerite ore sepctroscopically. They observed prominent indigo blue lines which were never noticed prior to this investigation. On the basis, of the indigo-blue line, the new element was named "indium".

Occurrence

Indium does not occur in the native state. In the combined state it occurs very widely in nature in small quantities. Its main source is Zinc blende in which it occurs to the extent of 0.1 percent. Indium almost occurs invariably in minute amount in cassiterite (tin stone), and also often found in pyrites. In traces, it also occurs in smithsonite and calamine. Lockyer reported the presence of indium lines in the solar spectrum.

EXTRACTION

There are number of methods for the extraction of indium from various sources. Some of the important ones can be discussed as follows:

1. Reich and Richter's Method from Zinc Blende

Zinc blende is powdered and dissolved in nitric acid. Hydrogen sulphide gas is allowed to pass through the solution, when heavy metals like Cu, Pb, As, Cd and Mo are precipitated which are filtered out. The filtrate is treated with an excess of ammonia solution when most of the zinc goes into solution while indium and some other hydroxides are precipitated.

The precipitate is first of all washed and then dissolved in acetic acid and hydrogen sulphide is passed into the solution, which precipitates indium sulphide and also the sulphides of Fe, Zn and Mn etc. The ppt is now again dissolved in hydrochloric acid and reprecipitated with an excess of ammonium hydroxide. This is repeated so as to remove completely zinc and manganese.

The final precipitate of indium hydroxide is dissolved in acetic acid and treated with hydrogen sulphide, when pure beautiful yellow precipitate of indium sulphide is obtained.

$$In(OH)_3 + 3CH_3COOH \rightarrow In(CH_3COO)_3 + 3H_2O$$

$$2In(CH_3COO)_3 + 3H_2S \rightarrow In_2S_3 + 6CH_3COOH$$

2. Winkler's Method

The indium having zinc is treated with dilute sulphuric acid insufficient to dissolve all the metal. On the residue of undissolved zinc, a spongy mass of lead, copper, tin, arsenic, iron and indium gets deposited. The spongy material is collected and mixed with strong sulphuric acid and heated strongly so as to volatilize most of the acid. To this solution an excess of ammonia is added which precipitates indium hydroxide along with the hydroxides of Fe, Zn and cadmium.

This precipitate is washed and made to dissolve in a little hydrochloric acid and the ferric iron is reduced to ferrous by sulphur dioxide. The solution is now treated with barium carbonate when indium gets precipitated as hydroxide.

Metallurgy

1. Indium can be obtained by reducing its oxide with hydrogen, carbon or coal gas (Reich and Richter),

$$In_2O_3 + 3H_2 \rightarrow 2In + 3H_2O$$

$$In_2O_3 + 3C \rightarrow 2In + 3CO$$

2. It is also obtained by Winkler's method by heating the oxide or chloride of indium with metallic sodium under a layer of dry sodium chloride.

$$In_2O_3 + 6Na \rightarrow 2In + 3Na_2O$$

$$InCl_3 + 3Na \rightarrow 2In + 3Na$$

3. Rossler and Wolf obtained indium by the addition of zinc shavings to a slightly acid solution of an indium salt ($InCl_3$), when metallic indium is precipitated.

$$2InCl_3 + 3Zn \rightarrow 2In + 3ZnCl_2$$

4. Schucht obtained indium by the electrolysis of indium sulphate or chloride or nitrate in the presence of pyridine, hydroxylamine or formic acid.

PROPERTIES

Physical

Indium occurs as a silvery white metal like that of platinum. The electro-deposited metal forms octahedral crystals belonging to cubic system, the crystals being isomorphous with gallium and aluminum. As indium is a very soft metal, it readily marks paper. It is a ductile metal. It melts at 155°C. Indium salts impart bluish-red colour to the flame.

Chemical Action of Air

At ordinary temperature, indium is quite stable in air, and retains it lusture. When it is heated above its melting point, a film of monoxide, InO is formed and on heating further at higher temperature it yields indium trioxide, In_2O_3.

Action of halogens : It directly reacts with halogens halides.

Action with sulphur : With sulphur on heating, the combination occurs with incandescence.

Action of acids : It dissolves slowly in cold dilute mineral acids, but more rapidly when heated, with the evolution of hydrogen.

$$2In + 6HCl \rightarrow 2InCl_3 + 3H_2$$

Indium dissolves in a solution of oxalic acid but not in acetic acid

Action of alkalies : Indium is not effected on treating with potassium hydroxide.

Action with other metals : indium readily alloys with sodium gold, tin, lead gallium and platinum.

Uses

Indium does not find much use. It is having the characteristic to harden to provide strength to the metal with which it is alloyed. The alloys of indium are highly resistant to corrosion and do not tarnish easily and thus find use in jewellery and dental work. Indium compounds are used for colouring glass, paints, fluorescent lighting etc.

INDIUM COMPOUNDS

Oxides of Indium. Indium is known to form number of oxides. Its hemioxide, In_2O is a black coloured solid. It is obtained by reducing the trioxide, *i.e.,* by heating the trioxide in hydrogen at 500°C. followed by sublimation in a high vacuum at about 700°C. It acts as a powerful reducing agent and liberates hydrogen from acids. On heating in air it is readily oxidized into sesquioxde.

Indium monoxide, In_2O_3 is the stable oxide. It is formed if the metal burns in oxygen, or if the hydroxide or a salt with a volatile acid is ignited. Indium sesquioxide is yellow at room temperature, but brown when hot.

Halides of indium : Indium is known to form number of halides such as fluorides, chlorides, bromides and iodides, but its most important halide indium trichloride $InCl_3$ is described here.

Indium Trichloride, $InCl_3$: Preparation

1. It is prepared by heating indium dichloride in a current of chlorine.
2. It is also obtained by heating the trioxide and carbon mixture in a current of chlorine.

   $In_2O_3 + 3C + 3Cl_2 \rightarrow 2InCl_3 + 3CO$
3. Another method of its preparation is by heating the metal in moist hydrochloric acid gas at 120-130°C.

Properties

It exists as colourless crystalline substance. It melts at 586°C into a pale-yellow liquid. Its Sp. Gr. is 4.0. It is a deliquescent substance and dissolves in water. If its aqueous solution is evaporated at a high temperature, a white powder of indium oxychloride, InOCl is obtained. Indium chloride acts as a catalytic agent in chlorinating benzene.

Indium sulphate, $In_2(SO_4)_3.9H_2O$: It is obtained by dissolving the oxide, hydroxide, or carbonate of indium in sulphuric acid, and evaporating the solution to almost to a syrup.

This is then cooled in a desiccator when deliquescent crystals of hydrosulphate, are obtained. When dried at 100°C, the sulphate $In_2(SO_4)_3.9H_2O$ is obtained as a white powder. It exists as a white powder. It also forms double sulphates and alums with ammonium sulphate, rubidium sulphate etc.

$(NH_4)2SO_4.In_2(SO_4)_3.In_2(SO_4)3.24H_2O; Rb_2So_4.In_2(SO_4)_2.24H_2O$

Analytical Reaction

1. Indium salts with sodium or ammonium carbonate yield a gelatinous precipitate of indium carbonate. This precipitate of indium carbonate is soluble in excess of ammonium of ammonium carbonate but reprecipitates on boiling.
2. With oxalic acid and alkali oxalates, indium salts yield crystalline precipitation of indium oxalate.
3. Indium salts yield a whit precipitate with potassium ferrocyanide and a yellow precipitate with potassium chromates.
4. Quantitatively, indium salts can be precipitated with organic bases like dimethyl aniline, piperidine, quinoline etc.
5. Indium salts yields white precipitate of indium hydroxide with a solution of hydroxylamine.
6. Indium can also be detected by using either alizarin or quinalizarin. One or two drops of neutral or slightly acidic indium salt solution is kept on a filter paper which is previously impregnated with a saturated alcoholic solution of alizarin or quinalizarin. The moist paper is kept over ammonia and then finally immersed in a saturated queous solution of boric acid. Red or violet against the yellow background of paper, reveals the presence of indium.

Estimation

Quantitatively it could be estimated by precipitating with ammonia as hydroxide and would be weighed as oxide after subsequent ignition.

RHODIUM

This element was discovered by *Dr. W. H. Wollaston* in 1803. He got a beautiful dark red powder after the removal of platinum form the crude platinum metal. This powder obtained was later on found to be a double fluoride of sodium and a new metal. On account of the rose colour of its salt, Wollaston named the new metal as rhodium.

Occurrence

Rhodium occurs in association with platinum ores in osmiridium and also in rhodite from Brazil. In very small amounts it also occurs in earth's crust.

Extraction

Woolaston's Process : The crude platinum metal was first of all dissolved in aqua-regia and then neutralized the excess acid with caustic soda. To this ammonium chloride was added to precipitate platinum as ammonium chloroplatinate. In the filtrate mercury cyanide was added to precipitate palladium as palladous cyanide. The filtrate was evaporated with sodium chloride to dryness and freed from mercurous cyanide and the residue washed with 84% alcohol in which soluble iridium, platinum, and copper etc. got dissolved. A beautiful dark red powder of sodium chlorohodate was obtained. This when treated with zinc metal yields black flaky precipitate of rhodium metal.